# ASBESTOS: HEALTH RISKS AND THEIR PREVENTION

MEETING OF EXPERTS ON THE SAFE USE OF ASBESTOS

GENEVA, 11-18 DECEMBER 1973

INTERNATIONAL LABOUR OFFICE - GENEVA

ISBN 92-2-101229-8

*First published 1974*

The designations employed in ILO publications, which are in conformity with United Nations practice, and the presentation of material therein do not imply the expression of any opinion whatsoever on the part of the International Labour Office concerning the legal status of any country or territory or of its authorities, or concerning the delimitation of its frontiers.
The responsibility for opinions expressed in signed articles, studies and other contributions rests solely with their authors, and publication does not constitute an endorsement by the International Labour Office of the opinions expressed in them.

ILO publications can be obtained through major booksellers or ILO local offices in many countries, or direct from ILO Publications, International Labour Office, CH-1211 Geneva 22, Switzerland. A catalogue or list of new publications will be sent free of charge from the above address.

## Contents

|  | Page |
|---|---|
| Report of the Meeting of Experts on the Safe Use of Asbestos | 1 |
| Health Hazards of Asbestos by the Office of the Chief Medical Officer, Department of Employment, United Kingdom | 19 |
|    1. The Effects of Asbestos on the Human Body | 19 |
|    2. Factors Influencing the Harmful Effects of Asbestos | 21 |
|    3. Prevention of Disease | 22 |
|    4. Monitoring the Health of Workers | 23 |
|    5. The Department of Employment Survey of Asbestos Workers | 24 |
| Pathology of Asbestos by Professor J. Champeix | 29 |
|    1. Introduction | 29 |
|    2. Criteria for Detecting Asbestosis | 29 |
|    3. Problems Presented by Exposure to Asbestos Fibres and the Appearance of a Pleural Mesothelioma | 33 |
|    4. Bronchial Cancer and Asbestos | 66 |
| Technical Prevention of Asbestos Hazards by A. Wilkie | 72 |
|    1. Assessment of the Exposure | 72 |
|    2. Methods of Protection | 75 |
|    3. Problem Areas Remaining | 80 |
| Survey of Statutory Provisions relating to the Prevention of Health Risks due to Asbestos | 81 |
|    1. Introduction | 81 |
|    2. General | 81 |
|    3. Scope of the Problem | 82 |
|    4. National Technical Provisions relating to Asbestos | 85 |
|    5. Medical Supervision | 91 |
|    6. Information and Training | 93 |

## REPORT OF THE MEETING OF EXPERTS ON THE SAFE USE OF ASBESTOS

(Geneva, 11-18 December 1973)

1. Following a decision taken by the Governing Body of the ILO at its 188th Session (Geneva, November 1972) a Meeting of Experts on the Safe Use of Asbestos was held in Geneva from 11 to 18 December 1973.

2. The agenda of the meeting was as follows:

(1) Pathological effects of exposure to asbestos (including asbestosis and cancer).

(2) Prevention of risks due to exposure to asbestos:

    (a) technical prevention;

    (b) medical prevention;

    (c) administrative measures.

(3) Possibilities of international regulations.

3. The following took part in the meeting:

Dr. Erik BOLINDER,
Medical Adviser to the Swedish Trade
  Union Confederation,
Barnhusgatan 18,
10553 STOCKHOLM   (Sweden)

Mr. Bernd W. von BRAUCHITSCH,
Chairman of the Management of Jurid-Werke GmbH,
  and of the Wirtschafts Verband Asbest,
Postfach 6,
2057 RHEINBEK   (Federal Republic of Germany)

Dr. A. CAVIGNEAUX,
Médecin-conseil de l'Institut national de
  recherche et de sécurité,
30 rue Olivier Noyer,
75014 PARIS   (France)

    Technical Adviser:   Prof. Jean CHAMPEIX,
                       Faculté de Médecine,
                       28 Place Henri-Dunant,
                       63001 CLERMONT-FERRAND (Cedex)   (France)

Mr. A.A. CROSS,
Chairman,
Environmental Control Committee,
Asbestosis Research Council,
114 Park Street,
LONDON, W1Y 4AB   (United Kingdom)

        Technical Adviser: Dr. W.J. SMITHER,
                         Medical Consultant to the Asbestosis
                           Research Council,
                         114 Park Street,
                         LONDON, W1Y 4AB   (United Kingdom)

Mr. L. EATON,
Regional Secretary,
Union of Construction, Allied Trades
  and Technicians,
293/295 Kentish Town Road,
LONDON, N.W.5.   (United Kingdom)

        Technical Adviser: Dr. R. MURRAY,
                         Medical Adviser to the Trade Union
                         Congress,
                         Congress House,
                         Great Russell Street,
                         LONDON, WC1B 3LS   (United Kingdom)

Mr. W. JOHNSEN,
Manager of Dansk Eternit-Fabrik A/S,
P.O. Box 763,
DK-91000 AALBORG   (Denmark)

Dr. W. KLOSTERKÖTTER,
Director,
Institute of Occupational Health and
  Industrial Hygiene,
55 Hufelandstrasse,
43-ESSEN   (Federal Republic of Germany)

Dr. S.F. McCULLAGH,
Chief Medical Officer,
James Hardie and Coy. Pty. Ltd.,
P.O. Box 219,
GRANVILLE, N.S.W. 2142   (Australia)

        Technical Adviser: Mr. R.D. PALFREYMAN,
                         Director,
                         James Hardie and Coy. Pty. Ltd.,
                         P.O. Box 219,
                         GRANVILLE, N.S.W. 2142   (Australia)

Dr. J.C. McNULTY,
Director,
Occupational Health Division,
Public Health Department,
57 Murray Street,
PERTH, W.A. 6000   (Australia)

Mr. George PERKEL,
Research Director,
Textile Workers Union of America,
99 University Place,
NEW YORK, N.Y. 10003   (USA)

Mr. N. PON,
International Organiser,
International Association of Heat and Frost
  Insulators and Asbestos Workers,
13220 - 79th Street,
EDMONTON, Alberta    (Canada)

Mr. B.B. TAYLOR,
Director,
Asbestos Cement Limited,
P.O. Box No. 486,
6 South Leinster Street,
DUBLIN 2    (Ireland)

    Technical Adviser:  Dr. S. HOLMES,
                                Secretary,
                                Asbestosis Research Council,
                                Turner Bros. Asbestos Co. Ltd.,
                                P.O. Box 40,
                                ROCHDALE, Lancashire    (United Kingdom)

Mr. A.G. WILKIE,
H.M. Chemical Inspector of Factories,
Department of Employment,
403-405 Edgware Road,
LONDON, N.W.2.    (United Kingdom)

Dr. R. Bonsanti, of the Italian Federation of Woodworkers, Construction Workers and Allied Trades (CGIL), invited to the meeting, at the last moment could not attend.

International Organisations

World Health Organization

Dr. G.E. LAMBERT,
Scientist,
Occupational Health Unit,
World Health Organization,
GENEVA    (Switzerland)

International Agency for Research on Cancer

Dr. P. BOGOVSKI,
Chief,
Unit of Environmental Carcinogens,
International Agency for Research on Cancer,
150 Cours Albert Thomas,
69008 LYON    (France)

    Technical Adviser:  Dr. J.C. GILSON,
                                Director,
                                MRC Pneumoconiosis Unit,
                                Llandough Hospital,
                                PENARTH. CF6 1XW
                                Glamorgan    (United Kingdom)

Council of Europe

Mr. L. HERTWIG,
Administrateur,
Affaires Economiques et Sociales,
67006 STRASBOURG Cedex    (France)

Commission of the European Communities

Dr. P. HENTZ,
Chef de la division Médecine et
  hygiène du travail,
Direction générale des Affaires sociales,
LUXEMBOURG    (Grand-Duché de Luxembourg)


Non-Governmental Organisations

International Organisation of Employers

M. JOIN,
Délégue Général de la Chambre Syndicale
  de l'Amiante,
10 rue de la Pépinière,
75008 PARIS    (France)

Dr. Jaques LEPOUTRE,
Médecin en chef des services médicaux d'Eternit S.A.,
Secrétaire du "Comité d'information d'amiante",
9 rue Ducole,
1000 BRUSSELS    (Belgium)

M. Emidio ANGELLOTTI,
Directeur de la Società Amiantifera
  di Balangero,
10070 BALANGERO    (Italy)

Permanent Commission and International
Association on Occupational Health

Dr. P.V. PELNAR,
Scientific Secretary,
Institute of Occupational and
  Environmental Health,
Suite 412,
5 Place Ville Marie,
MONTREAL 113    (Canada)

4.    Mr. Astapenko, Assistant Director-General of the ILO, opened the meeting. He recalled the work of the ILO in relation to problems of health protection of workers exposed to airborne dust, and in particular the action carried on with regard to pneumoconiosis. The growing evidence of severe health risks following exposure to asbestos dust has raised great concern in the ILO Governing Body, which felt that effective action should be taken for keeping such risks under control. The meeting of experts was convened in order to give advice on the appropriate preventive and control measures to be taken, and on the type of action the ILO could take.

5. The meeting unanimously elected Dr. J.C. McNulty as Chairman and Mr. A.A. Cross and Dr. R. Murray as Reporters.

## Definition of asbestos

6. Asbestos is a broad term applied to a number of substances falling into two chief varieties, chrysotile and the amphiboles. These substances are naturally-occurring iron, sodium, calcium and magnesium hydrated silicates; they have a fibrous structure and are incombustible. Chrysotile asbestos (white asbestos) is a hydrated magnesium silicate found in serpentine rock. It is widely distributed in nature and accounts for some 93 per cent of the world's asbestos production. Amphibole asbestos varieties include amosite, crocidolite, anthophyllite, tremolite and actinolite. The last two substances have few industrial applications but are sometimes found as impurities in talc.

## Pathological effects of exposure to asbestos

7. An understanding of the biological effects of asbestos is essential to its safer use. The risks to health result from the inhalation of the fibrous dust and its subsequent dispersion within the lung and to other parts of the body. This may occur in those engaged in extracting and processing the fibre, and subsequent manufacture and application of products. In practice, exposure to asbestos dust alone is uncommon; other mineral dusts are frequently inhaled along with the asbestos and may influence its effects. The effect of these and other pollutants, especially cigarette smoke, may adversely influence the type and severity of disease produced by asbestos.

8. Several types of disease may result from inhalation of asbestos:

(a) Asbestosis - a fibrosis (thickening and scarring) of the lung itself and of its outer surface, the pleura which may become calcified.

(b) Cancer of the bronchial tubes.

(c) Cancer of the pleural surface (diffuse mesotheliomas). Diffuse mesotheliomas may also occur in the abdominal cavity (peritoneal mesotheliomas).

(d) There is also some evidence that cancers in other sites of the body may occasionally be asbestos-linked.

These types of disease are first manifest years or even tens of years after first exposure to the dust so that the recent rising incidence of cases in many countries is the result of past dust exposures which were clearly far too high. Current incidence of disease is therefore not a measure of the effects of present dust levels.

9. Asbestosis develops slowly and is detectable by a combination of clinical, radiographic and lung function tests. In the early stages there will be uncertainty about diagnosis. The ILO U/C 1971 International Classification of Radiographs of Pneumoconioses

is of value in grading the continuum of radiographic changes. An agreed international grading of severity taking into account all diagnostic features might be of value in improving comparability of statistics of disease incidence. Once the disease has reached a stage where diagnosis is certain, the fibrosis tends to increase despite removal from further dust exposure. It is not yet known if there are detectable early stages at which removal from further dust exposure markedly influences the course of the disease.

10. The bronchial cancers associated with asbestos are not distinguishable from those produced by cigarette smoking and other causes, but there is evidence of synergism between cigarette smoking and asbestos so that the risk of developing bronchial cancer is materially greater in asbestos workers who smoke cigarettes than in the general public with similar smoking habits. There is evidence that where dust conditions have been improved and the incidence of asbestosis greatly reduced, the excess bronchial cancer risk has also fallen.

11. The incidence of mesotheliomas is less closely related to the risks of asbestosis and are probably unrelated to cigarette smoking. They may occur many years after even a short exposure to asbestos but their incidence is in some measure dose-related; those with the heaviest past exposures having a higher (or earlier) risk of developing the tumours.

12. Mesotheliomas are an exceptionally rare form of cancer in the general population. Most of the cases are apparently related to past exposure to asbestos, but in most surveys a minority have apparently had no such association. With the probable exception of anthophyllite, all types of fibre are associated with mesotheliomas but there is evidence that the risk may be greater with crocidolite than with amosite or chrysotile. The proportion of asbestos workers who may be expected to develop mesotheliomas cannot be determined precisely at present because of the very long lapse period - over fifty years in some instances - between first exposure and the occurrence of a tumour. Present evidence indicates that the risk was highest in insulation workers heavily exposed in the past. In this special section of the industry, the proportion may have been in the order of 10 per cent. This compares with over 80 per cent in the case of some potent chemical carcinogens such as beta-naphthylamine.

13. The risk of asbestosis and of asbestos-related cancers has been studied in many defined groups of asbestos workers. No simple pattern of risk related to type of job, type of fibre, past dust exposure or other factors has emerged. This may in part be due to inadequacy of past information, but also to differences in the physical and chemical characteristics of the dust which influence both its access to the lungs and to the cells responding to the dust. Exposures to a single type of fibre have occurred principally during mining and initial processing. Exposures during manufacture and in insulation work have usually been to more than one type of fibre and to other minerals.

14. The highest excess risks of asbestos-related cancers and asbestosis have been in the insulation workers (laggers). In those asbestos textile groups where the incidence of asbestosis was very high forty years ago, and where improvement in dust conditions first occurred, asbestos-related diseases have been greatly reduced. The

incidence of disease in, for example, the asbestos cement and friction materials section of the industry has, in general, been lower than in those parts of the industry where the fibres are less firmly bound in the products. In chrysotile mining and milling, despite very heavy dust concentrations in the past, the incidence of severe asbestosis, asbestos-related cancers and especially mesotheliomas has been low.

15. In conclusion[1]:

(a) It is well established that exposures to asbestos dust of all types have caused serious disease.

(b) There is evidence that the incidence of diseases is related to the intensity and duration of exposure. Thus it may be possible to determine levels of dustiness where the risk may be minimised to an acceptable level.

(c) Estimates of what these levels may be have been made using some of the available evidence. A more comprehensive review of all the evidence is still required, as well as further research including prospective studies.

(d) Much is not yet understood about factors influencing the incidence and severity of asbestos-related diseases but it is unnecessary to wait for a complete understanding or overwhelming proof of one point to make the maximum use of existing knowledge to reduce risks to a minimum in the future.

## Prevention of Risks Due to Exposure to Asbestos

### (a) Technical Prevention

16. Having regard to the risks indicated in the foregoing review of the pathological effects of exposure to asbestos dust, the experts have been asked to consider methods whereby such risks can be prevented. The problem of the prevention of harmful dust emission is complicated because of the wide and varied range of products containing asbestos and the variety of uses and applications. The peculiar and, in some respects, unique properties of asbestos fibres have led to their utilisation in combination with many other materials. It is said that there are more than 3,000 different varieties of such products, varying from the familiar asbestos cement corrugated sheet to the very sophisticated asbestos-reinforced material used in heat shields on spacecraft.

17. This variety of products requires a considerable diversity of processes for their manufacture; the properties of resis-

---

[1] Summaries of current knowledge about the effects of asbestos are available in IARC Monograph of the Evaluation of Carcinogenic Risk of Chemicals to Man. IARC 1973, in the Proceedings of the IARC Symposium on the Biological Effects of Asbestos, Lyon, 1972 (IARC 1973), and the Report of the Advisory Committee on Asbestos Cancers to the Director of the IARC (1973).

tance to heat and corrosion of asbestos fibres make them virtually indestructible. Thus in many applications the materials will endure to the point that when they must eventually be removed this should be done in such a manner that, as in the case of manufacture and application, the possibility of harmful dust emission is prevented.

18. Safer substitute materials rendering the same purpose as asbestos are desirable and in some instances apparently less harmful materials have been developed. Such substitutes should be used whenever possible, but they should be subjected to the appropriate tests to determine their relative safety, fire and health risks. Nevertheless, it is apparent that the properties of asbestos fibre are such that its continued use is unavoidable in a number of materials.

19. Even where alternative materials have replaced asbestos, there will for many years be installations - plant, buildings, ships - where in due course the asbestos material formerly used (principally in the form of thermal insulation) has to be removed. For many years, therefore, it will be necessary to ensure that such work is performed in such a way that the health of men engaged in this work is not impaired.

20. Consideration of the techniques by which people can be protected from risks arising from asbestos indicate that a fundamental requirement is to distinguish between, on the one hand, those situations where the form of the product and/or the manner in which it is used makes it difficult or impossible to prevent the emission of harmful quantities of asbestos dust and, on the other hand, those materials where the asbestos is so bonded with other materials that in normal use there is little possibility of harmful emission.

21. It follows that some means of measuring the dust exposure is essential for the guidance of those devising techniques of control, and for the monitoring of the environment to ensure that required standards are being achieved and maintained. Nevertheless, the experts stress that whatever standards of maximum dust exposure may be recommended as targets, the ultimate aim will always be the minimum possible amount of occupational exposure.

Standards of concentrations of asbestos dust

22. With any known carcinogen, all exposures must be minimised, but for practical purposes it is desirable that the competent authority establishes target standards, methods of sampling, measurement and frequency of monitoring. It is recommended by the experts that the membrane sampling method[1] as described by the Asbestosis Research Council should be adopted as an international reference method for comparison and correlation of environmental data concerning asbestos dust concentrations.

23. In order to establish more confidently a safe standard for all types of asbestos-related disease, much additional data are needed. The experts recommend that, in the present state of knowledge, the 2 fibres/cm$^3$ time-weighted standard which had been adopted by some member States should be regarded as an interim

---

[1] Asbestosis Research Council Technical Notes 1, 2 and 3.

target concentration for the prevention of risk to the health of asbestos workers. It was recognised that this standard related to the fibrogenic effects and not to the carcinogenic effects for which no standards exist at the present time.

## Technical Prevention of Harmful Concentrations of Asbestos Dust

### Dust suppression

24. Wherever technically possible, the process or operation should be designed or redesigned so that dust exposure is reduced to the recommended target concentration or even lower. For example, this can sometimes be accomplished by suppression through wetting or incorporation of the asbestos component into other dust-suppressing materials or compounds. Other examples of control effectively achieved by such methods can be found in certain textile processes by moistening the asbestos which is being processed. Thorough wetting by spray nozzles inserted into old lagging substantially reduces the dust produced during the removal of the asbestos material. Certain asbestos compositions for insultation can be wetted in their bags before being extracted and applied.

25. Such considerations apply to processes from which dust would be evolved unless such measures were taken to prevent it, but there are many operations with asbestos containing materials which are intrinsically dust free. For example, the punching out of rubberised asbestos gaskets is a process which does not produce dangerous dust concentrations because the bonding effect of the rubber reduces the escape of fibres produced by the operation into the atmosphere in the form of inhalable asbestos dust. It has been found possible to manufacture some asbestos products (e.g. textiles, millboard) in a dust-suppressed form so that they can be used in many applications without producing harmful quantities of dust.

### Enclosure and mechanisation

26. Where it is not possible so to modify a process that asbestos dust is not emitted, the first consideration should be to localise or enclose that process or the dust-producing part of the process so that dust does not escape into the workplace. Wherever possible, mechanical means should be devised to avoid handling involving personal exposure.

27. Wherever possible, manufacturers of equipment specifically intended for use with asbestos should be required so to design and construct the equipment that it does not permit the escape of dust.

### Partial enclosure with exhaust ventilation (dust extraction)

28. Where total enclosure is not practicable, in many cases partial enclosure together with exhaust ventilation can be used.

Even with totally-enclosed systems, sufficient exhaust ventilation is necessary to provide negative pressure and avoid dust leakage from joints in the system.

29. The nature of the process will determine the extent to which the process or operation can be enclosed. Openings in enclosures should be as small as possible consistent with access to the work and safe manipulation carried out inside the enclosure. Exhaust draught should be applied to the interior of the enclosure so that air moves inwards through the opening at sufficient speed to ensure that the breathing zone of the operative is free from dust and that dust does not escape into the workplace. In some cases, access to the work may be such that a large opening is necessary. Such an enclosure may be more accurately described as a booth. In such cases, particular care is necessary in the design. The capacity of the booth, and the capacity, air speed and positioning of the ventilation equipment have to ensure that turbulence and eddies created by operatives and mechanical movement etc. do not obstruct the effective removal of respirable dust from the breathing zone of the operative and the workplace in general.

Receptor and captor hoods

30. When the process cannot be enclosed in the manner described in the preceding paragraphs, hoods, forming part of an exhaust ventilation system, should be fitted as near as possible to the source of the dust. These exhaust systems may be designed to receive and remove the dust-laden air that the process is delivering to them, or they may be designed to deflect dust particles that are moving in a direction imposed on them by the process, and extract them before they pollute the general atmosphere.

31. Exhaust systems associated with receptor hoods must be capable of removing all the air that the process is delivering to the hood, whereas the fan of a captor system must draw air into the hood at a speed that is capable of changing the direction of the dusty air from the process. Systems based on "low-volume and high-velocity" principles have proved effective in coping with operations where high velocities are involved (e.g. abrasive wheels) and should be considered if the methods described in the preceding two sections are impracticable. It should be emphasised, however, that the design of extraction systems for any particular process requires professional expertise and it should therefore be referred to a ventilating engineer of sound professional standing.

32. In many areas of use, for example, on building sites, it may be necessary to use mobile or portable equipment. Tools fitted with specially-designed hoods have been developed for use with mobile dust collection units and, if approved as suitable for asbestos by the competent authority, should be used.

Control of dust collected in
ventilation systems

33. Dust collection by exhaust draught requires at some stage the separation of the asbestos dust from the air before it is readmitted to the atmosphere. It is important that the filtration of the air should be so effective and reliable that harmful

emissions of asbestos dust are not transmitted to the general environment or returned to the workplace. Filtered air must not be readmitted to the workplace unless asbestos dust levels are not more than one-tenth of the recommended target concentration for the working environment.

34. Filtration equipment must be regularly inspected and serviced to ensure that it continues to operate at the designed efficiency. Asbestos dust collected by the unit must be regularly removed under strictly-controlled hygienic conditions and removed for disposal in impermeable containers as described in the section of this report dealing with waste disposal.

## Maintenance of technical means of dust control

35. All types of dust control equipment should be inspected by technically-competent persons under conditions specified by the competent authority. A certificate of effectiveness should be issued and made available for inspection. It should also be the responsibility of the employers to ensure that such equipment is kept under supervision to ensure that it is at all times operating in the manner intended.

## Personal protection

36. Where it is impracticable by technical measures to prevent the possibility of inhalation of asbestos dust, personal protection must be provided in the form of respiratory protective equipment and protective clothing. The use of such equipment should be limited to the interim period required to achieve dust control by technical measures required by the competent authority or to those situations where adequate technical control of dust is impracticable.

37. The type of respiratory protective equipment appropriate for different levels of asbestos dust concentrations should be specified by the competent authority, which should also take responsibility for establishing standards and methods for approval of such equipment for use with asbestos. Well-fitted respiratory protective equipment of the half-face mask will generally be found to be suitable for concentrations up to twenty times the target concentration. Where concentrations could exceed such a level (for example, in some cases of filtration plant maintenance or removing old asbestos lagging) more sophisticated equipment, such as positive-pressure respirators (where the fresh air supply is fed by a small portable electric pump through a filter into the face piece), would be required. In extreme cases, direct air-line breathing apparatus might be considered necessary.

38. Some experts considered that there was a place for the use of lightweight disposable face masks of a type approved by the competent authority. Others were apprehensive that the greater comfort of this type of mask might encourage their use in concentrations for which they were not adequate, engendering a false sense of security. It was recognised that the development of effective half-face respirators which could be worn with less discomfort, especially in warm weather or warmer climates, was very desirable.

39. It is essential that persons who are required to wear protective respiratory equipment should be individually fitted and properly trained in its use, and should be fully aware of the reasons why and occasions when the equipment must be worn. In view of the discomfort of wearing half-face respirators, the duration of continuous use should be limited.

40. Respiratory equipment should be issued to individuals for their exclusive personal use and exchanged at regular and suitable intervals for cleaning and maintenance by suitably-qualified personnel. Equipment when not in use should be stored in suitably-hygienic containers so that it does not become contaminated.

41. Protective clothing should be designed to prevent the deposition of asbestos dust on the personal clothing of the wearer. It should be issued for the personal use of the wearer and exchanged for cleaning at such intervals as conditions of use necessitate.

42. Accommodation should be provided for changing into and out of protective clothing at the beginning and the end of each work period and for the accommodation, free from the possibility of contamination, of employees' personal clothing not worn at work.

43. Facilities, such as vacuum equipment, should be provided for the removal, at the end of the work period, of dust deposited on the protective clothing. On no account should dust be removed by compressed air or brushing. Protective clothing should not be worn outside the area or place of work for which it is provided.

44. Protective clothing collected for cleaning should be packaged in impermeable containers, such as plastic sacks, labelled as asbestos-contaminated clothing. Cleaning should be carried out under suitable arrangements (e.g. by wetting before unpacking) by cleaning services which understand the need to ensure that their own staff are not exposed to a risk from asbestos dust.

General hygiene

45. When asbestos is being used, particular attention should be paid to the regular cleaning of premises and plant so that deposits of dust do not accumulate.

46. Cleaning should be carried out by vacuum cleaning equipment which has been approved for use with asbestos by the competent authority. Alternatively, cleaning can be carried out by wet methods so as to prevent dust from being given off into the atmosphere of the working environment. Should it not be possible to use such equipment or other equally effective methods, workers engaged in cleaning should be provided with approved personal protective equipment. Cleaning should be carried out in such a manner or at such times that other workers are not exposed to risk, or the area being cleaned should be screened to prevent contamination of adjacent areas.

47. Subject to conditions laid down by the competent authority, the employer should be responsible for ensuring that asbestos dust given off in the place of work or removed in the form of waste does not adversely affect other parts of the factory or the general environment.

## Demolition of plant buildings and shipbreaking where asbestos materials are involved

48. It is agreed by the experts that probably the greatest problem in controlling harmful dust exposure occurs during the removal of old insulation and in the demolition of premises, plants and ships in which asbestos has been used. High concentrations of dust could occur, and control by ventilation is in most cases impracticable. The experts recommended therefore that particular care should be emphasised for the protection of workers in such situations and the following methods of control applied:

(a) personal protection of a high order;

(b) the use of specially-designed tools and techniques;

(c) dust suppression by wetting to the maximum extent possible;

(d) waste material collected, packaged and disposed of in accordance with the recommendations under the section entitled "Waste disposal"; and

(e) the segregation of areas where such work is carried out to prevent risk to the health of other workers not involved in the work of removal.

It is recommended that wherever possible asbestos material should be removed separately and at an early stage of the demolition, and that the use of specialised workers, experienced in the procedures outlined in the above paragraph, should be encouraged.

## Waste disposal

49. Process waste, filtered asbestos dust, dust or waste collected by vacuum or other cleaning, should be collected in containers (such as impermeable plastic bags) and sealed so that dust cannot escape. These containers should be removed and disposed of in accordance with procedures laid down by the competent authority.[1]

## Storage and distribution

50. Standards should be laid down to ensure that asbestos fibres, mixtures of fibrous materials, process waste and materials in transit should not emit asbestos dust during storage and distribution.

51. Impermeable bags and methods of packaging should be such as to prevent the escape of asbestos dust and to facilitate handling in such a manner that the risk of damage to the package and consequent spillage should be reduced to a minimum. Asbestos fibre should be transported in containers wherever possible and palletised to avoid the use of hooks in unloading. Bags should be marked with an internationally-recognisable symbol indicating that hooks should

---

[1] The ARC Recommended Code of Practice for Handling and Disposal of Asbestos Waste Material is an example of procedures adopted in one member State.

not be used. Those concerned with the transportation of asbestos fibre should be warned of the need to observe high standards of hygiene.

52. Containers of asbestos fibre, asbestos products of a dusty nature and dusty waste materials, including empty bags previously containing asbestos, should be clearly labelled in such a way that persons handling the material are informed of the nature of the contents and warned of the need for taking suitable precautions. Consideration should be given to the design of an internationally-recognisable warning symbol.

(b) **Medical Prevention**

53. In view of the uncertainties which still exist regarding the dose-response relationship in asbestos workers and the dynamic situation in which new methods and discoveries are continually emerging, the experts agreed that it is necessary to maintain a high standard of medical supervision to ensure the best protection for exposed workers. It will be necessary for some considerable time ahead to try to correlate dust exposure to different types of asbestos with the development of signs and symptoms.

54. It was therefore agreed that workers assigned to a job involving potential exposure to asbestos dust should be subject to medical supervision. Attention was drawn to the need to ensure effective supervision in itinerant workers, e.g. in the thermal insulation industry. The competent authority in each country should determine the most appropriate procedures for medical supervision.

55. The purpose of medical supervision is to detect, at the earliest possible stage, those signs of disorder which might be caused or aggravated by exposure to asbestos dust, to assess the effectiveness of preventive measures and to determine the long-term effects of exposure.

56. Ideally, such medical supervision should include:

(a) a medical examination before being assigned to asbestos work. The point was made that some workers may already be suffering from respiratory disease or insufficiency, and it was unwise to expose such persons to the additional hazard of asbestos exposure. There were differences of opinion on the methods of determining the contra-indications to employment with asbestos and its products, but it was accepted that the principle was valid and that the need for medical supervision increased the longer the worker was employed;

(b) periodic medical examinations at appropriate intervals. The periodicity varies from country to country and from one to three years, and depends in some cases on the nature and extent of the possible exposure. The advantage of periodic medical examinations is not only to assess the clinical status of the worker as compared with the previous examination, but to provide counselling, for example in relation to cigarette smoking, and to reinforce the need for co-operation in the maintenance and use of technical or personal protective equipment;

(c) post-employment medical examinations. It was generally recognised that, because of the insidious nature of the disease, medical supervision must continue after exposure had ceased. Apart from the continuing need to make a diagnosis at the earliest possible stage in the interests of the individual worker, it is also important for compensation and epidemiological purposes.

(d) certification of current fitness in a manner prescribed by the competent authority.

57. The nature and extent of the examinations outlined above will depend on local circumstances, but they should include:

(a) a personal and occupational history (including smoking habits);

(b) general clinical appraisal, including auscultation of the chest;

(c) chest radiography. The point was made that small films are not always good enough for an early diagnosis of asbestos effects. Full-size films are desirable;

(d) lung function test. Attention was drawn to the ILO booklet on this subject (O.S.H. Series No. 6);

(e) other tests at the discretion of the physician.

It was agreed that while asbestos bodies might be found in the sputum, these were not diagnostic of the disease and their detection was of little significance.

58. The results of medical examinations must be carefully recorded together with the worker's employment history, including data on dust measurements, so that assessments can be made of any developments in the clinical condition, correlation can be made with his dust exposure, advice can be given with regard to claims for compensation and epidemiological information can be derived for research purposes. The method of record-keeping may vary from one country to another, but it is essential that:

(a) strict confidentiality in relation to individuals should be observed and personal files should be kept under the seal of medical secrecy;

(b) records should be preserved regardless of changes of employment according to procedures laid down by the competent authority in each country so that they are available for ensuring regular health supervision and for epidemiological research.

(c) **Administrative Measures**

59. In order to put into effect the technical and medical measures outlined above, to ensure the education and training of all concerned and to achieve the high standards of collaboration between employers and workers which are necessary for the control of the hazard, a number of administrative measures have to be considered.

## Compulsory notification

60. In view of the widespread use of asbestos, the experts agreed that the competent authorities must be informed of all work involving exposure to asbestos dust so that the inspection services can take the necessary steps to ensure the application of the necessary technical and medical preventive measures already described. Recognising the difficulties of notifying all the circumstances in which exposure to asbestos dust may arise - reference was made particularly to demolition operations - the experts considered that there should be a notification procedure, if necessary limited to particular occupations, by which the employer would inform the competent authority of work involving exposure to asbestos dust.

## Exemptions

61. The competent authority may grant exemptions from any or all of these technical or medical recommendations for control, e.g. in circumstances where it can be shown that the risks of inhaling asbestos dust are negligible.

## Education and training

62. The experts attached great importance to the need for a full understanding of the nature of the asbestos hazard by all persons concerned, both inside and outside the undertaking. It is only in the light of this understanding that effective measures can be taken for the control of the hazard.

63. Attention was drawn particularly to the need for an authoritative booklet, suitably illustrated by photographs, line drawings or cartoons, covering the manifold applications of asbestos and describing the technical and medical preventive measures. Some countries have already produced such literature which could form a model for an international text. Other methods of instruction, such as films and film strips, were mentioned and particular reference was made to the value of the spoken word, either to individuals, e.g. in the course of a medical examination, or to groups of workers or line management.

64. Reference was also made to the need for training of:

(a) physicians, so that they can fully appreciate the significance of asbestos exposure in their patients;

(b) occupational hygienists and ventilating engineers, in the methods of measurement and control of asbestos dust;

(c) representatives of employers and workers in industries handling asbestos;

(d) civil servants in the official bodies whose duty it is to administer the legislative provisions relating to the use of asbestos.

The level of training provided will depend on circumstances. Those persons working within the industries using asbestos will in general require more detailed knowledge than those outside.

## Collaboration between employers and workers

65. The experts recognised the value of effective joint consultation which involves the full and frank disclosure of information relating to dust measurements, epidemiological studies, plant inspections, waste disposal and the like. Reference was made to the practice in some countries of ensuring this interchange of information.

66. It was agreed that, without minimising in any way the responsibility of the employer for proper and effective preventive action, his representatives should co-operate with those of the workers in taking practical steps for:

(a) consultation with a view to devising procedures to ensure that the requisite action is taken for the workers' protection against the hazards of asbestos dust at their place of work; and

(b) co-operation in programmes of prevention and supervision, all persons being encouraged to take part.

67. At the same time, it was recommended that workers, technicians and any other persons whose work directly or indirectly involves a risk of exposure to asbestos dust must abide by the prescribed safety rules and make proper use of prevention and protection equipment. This would include the wearing of personal sampling apparatus where this is considered necessary.

## Future ILO Action and Possibility of International Regulation

68. The experts recognised that there will be a continuing need for further research in order to clarify the pathogenesis of the lesions due to asbestos fibres, in particular the malignant changes both in the lungs and outside the lungs. The question of the relationship of exposure to various types of asbestos fibres and the incidence of mesotheliomas requires further explanation. The ILO should stimulate further action in this field in collaboration with other competent national and international bodies.

69. The experts underlined the fundamental importance of proper monitoring of the working environment and the urgent need for standardising dust sampling and measurement techniques. An important requisite is an acceptable target level of fibres in air on the bases of (a) a time-weighted average, and (b) peak exposure which should not be exceeded. The problem is to reconcile the levels relating to the fibrogenic effects and those relating to bronchial cancer and mesotheliomas. The Meeting recommended that the ILO should initiate appropriate research and convene in the future further meetings of consultants/experts at appropriate intervals, in order to examine these problems and either to establish such levels or to recommend how they can be established. Consideration should be given to the establishment of international reference laboratories for the standardisation of dust sampling and measurement techniques.

70. The experts noted the importance of standardising criteria for the diagnosis and prognosis of asbestos-induced

diseases. An agreed international grading of severity, taking into account all diagnostic features, would be of value in improving comparability of statistics of disease incidence. The experts recognised the great value of the ILO/UC International Classification of Radiographs of Pneumoconioses for improving understanding and international comparability in the reading of radiographs of asbestosis and recommended that continuing efforts should be made to improve the quality of their presentation.

71. The experts recommended that there is a need for the adoption of an international instrument or instruments on the safe use of asbestos at the earliest opportunity and recommend that in the meantime:

(a)     the ILO should circulate this report to the ILO member governments and through them to the employers' and workers' organisations in the countries concerned for information and action on their part, and to other interested bodies;

(b)     the ILO should prepare a guide on the safe use of asbestos in consultation with appropriate national and international bodies and should ensure that it is kept up to date.

## HEALTH HAZARDS OF ASBESTOS

by

the Office of the Chief Medical Officer
Department of Employment
United Kingdom

### 1. The Effects of Asbestos of the Human Body

Asbestos fibres enter the body by inhalation[1] and the following effects have been observed:

(a) <u>An interstitial fibrosis of the lung (asbestosis)</u>. As it develops this type of fibrosis restricts the capacity of the lung to expand and at the same time impedes the transfer of oxygen from the air cells (alveoli) to the blood in the pulmonary capillaries. At a later stage the flow of blood through the pulmonary vessels is impeded leading to cardiac strain and ultimately failure. Breathlessness on exertion is the characteristic symptom. In about 50 per cent of cases the defect in oxygenation of the blood is accompanied by changes in the nail beds of the fingers and toes (clubbing). The increased resistance to air-flow in the lungs produces characteristic crackling sounds during inspiration (rales or crepitations), which are audible via the stethoscope. Measurements of pulmonary ventilation demonstrate the reduced capacity of the lungs while the speed of expiration is unimpaired: the ratio of forced expiratory volume in one second to forced vital capacity ($FEV_1/FVC$) is thus increased - "restrictive defect". Measurements of oxygen uptake show the "gas transfer defect". The chest radiograph shows typical changes associated with interstitial fibrosis.

(b) <u>Pleural plaques</u>. Thickening of areas of the surface membrane of the lungs (the pleural) is found quite frequently in asbestos exposed persons. Asbestosis may or may not be present. The plaques often become calcified and are then visible on chest radiographs. In themselves they appear to cause no impairment of lung function. Whether they are of any significance in the development of pleural mesothelioma (see below) remains to be elucidated.

(c) <u>Lung cancer (carcinoma of bronchus or lung)</u>. An increased incidence of lung cancer in persons affected by asbestosis was reported by Doll [1955] and Buchanan [1963]. A less marked increase has also been reported in insulation workers exposed to asbestos who were not suffering from asbestosis [Selikoff, Churg and Hammond, 1964]. Because the latter series was not supported by autopsy evidence some authorities have felt some doubt that the lung cancer risk is truly increased in persons who are asbestos exposed but not suffering from asbestosis [Senior Medical Inspector of Factories' Advisory Panel, 1967]. The cause of cancer is unknown, but the likelihood of its

---

[1] The possibility of entry via the gastro-intestinal tract will be discussed later.

occurrence is known to be affected by several factors. Among these are inheritance and race; these are unlikely to influence the results of surveys of different industrial groups, but may cause differences when comparing the results of surveys in different countries. For lung cancer the influence of cigarette smoking is supported by overwhelming evidence [Royal College of Physicians, 1971]: it is thought that cigarette smoking may be a co-factor in the production of lung cancer in association with asbestos exposure [Selikoff, Hammond and Churg, 1968; Berry, Newhouse and Turok, 1972], and the effect appears to be multiplicative rather than merely additive [Doll, 1971]. Cancers associated with industrial agents may appear only after long latent periods: for example, in workers exposed to beta-naphthylamine the occurrence of bladder cancers was found to be associated with a mean latent period of 18 years, with standard deviation of 7 years, by Case and Hosker [1954]. No means of early detection of lung cancer exists which would be applicable to whole industrial groups on the scale of current asbestos exposure. Although the interval between relevant exposure and appearance of lung cancer may be many years, the methods at present available for detection (chest radiography, cytological examination of sputum) can only provide the diagnosis at a comparatively late stage, and the margin of time before treatment will no longer be effective is narrow.

(d) <u>Diffuse mesothelioma</u>. Mesothelioma is a form of cancer in which tumour growth originates in the pleura (pleural mesothelioma), or in some cases from the peritoneum, the outer surface membrane covering the abdominal organs (peritoneal mesothelioma). By comparison with lung cancer this is a rare condition. The association of mesothelioma with asbestos was first noted in the North West Cape Province of South Africa, where crocidolite is mined [Wagner, Sleggs and Marchand, 1960]. Subsequent reports have shown the close association of mesothelioma with crocidolite mining in South Africa, with very few being reported from amosite mining and none from chrysotile [Wagner, Gilson, Berry and Timbrell, 1971; Harington, Gilson and Wagner, 1971]. In manufactured uses crocidolite has also been shown to be important [Gilson, 1973], often in mixtures with amosite or chrysotile; to a much less extent to chrysotile alone [Manucuso and El-Attar, 1967] and to pure amosite exposure in insulation manufacture [Selikoff et al; quoted by Gilson, 1973]. No cases were found in anthophyllite exposure by Kiviluoto and Meurman [1970]. In spite of careful interrogation a history of asbestos exposure is not always forthcoming in affected persons, or may reveal only minor degrees of contact, possible remote in time [Newhouse and Thomson, 1965]. The extent to which this tumour, like all other cancers, is naturally occurring in the population is unknown. There is no means of early diagnosis, and successful treatment must be very rare but has been reported [Beverley, 1973].

(e) <u>The formation of asbestos bodies</u>. A proportion of asbestos fibres reaching the lung may be incorporated into a proteinaceous iron-containing capsule. These "bodies", also called ferruginous bodies, as they may be formed from other fibres than asbestos, have no harmful effect, and are probably protective in origin. They are found in large quantities in persons affected by asbestosis, and in most people with a history of exposure to asbestos. They have also been reported in autopsy studies of the lungs of population groups without known asbestos exposure.

## 2. Factors Influencing the Harmful Effects of Asbestos

To reach the alveoli of the lungs fibres must be in the respirable size range. This is defined as a particle from 5 to 100 µm in length with a length to breadth ratio of at least 3:1. The total dose absorbed depends both on the concentration of respirable fibres in the air and the duration of exposure. Differences of effect however are observed with different types of fibre (whether chrysotile, amosite, anthophyllite or crocidolite), and also depend on physical factors, such as the state of dispersion of fibres, their size, both as to diameter and length, and their shape. Other factors which must be taken into account are smoking habits and individual immunological status. Finally it must be remembered that in relation to lung cancer, and especially to mesothelioma, the long latent period means that the effects should be related to exposure conditions many years earlier. The same is true, to a lesser extent, in relation to asbestosis.

(a) Exposure factors. In asbestosis an interval of several years from first exposure normally elapses before evidence of the disease is apparent, and the disease appears to be limited to those persons who work in industries in which asbestos is extensively used. The increased risk of lung cancer in association with asbestos exposure has also been shown to be related to the degree of exposure [Newhouse, Berry, Wagner and Turok, 1972].

(b) Type of fibre. The importance of the type of fibre in relation to the development of mesothelioma has already been stressed. Recently a theory to explain the differences, based on the physical characteristics of the fibres, has been put forward [Gilson, 1973]. The fine straight fibres of crocidolite have a much greater chance of reaching the periphery of the lung and the pleura than other types. The known difference in mesothelioma risk associated with crocidolite from the North West Province compared with the Transvaal is found to be explicable in that the diameter of fibres from the latter area is three times larger than that of the former. Larger fibres, especially chrysotile and anthophyllite, are probably important in causing asbestosis.

(c) State of dispersion. A lower incidence of lung cancers in miners of chrysotile than that found in manufacture, especially of textiles, is probably due to the greater degree of dispersion of the fibres in the latter, producing more dust in the respirable range [Gilson, 1973].

(d) Other factors. The great importance of cigarette smoking in relation to lung cancer and asbestos exposure has been mentioned above. There is no evidence that smoking has any influence on the occurrence of mesothelioma. The possibility that damage to the mucous membrane of the bronchial tubes from smoking may facilitate penetration of the larger fibres and increase the likelihood of asbestos developing is worthy of consideration. Differences in certain factors in the blood (rheumatoid factors and anti-nuclear factor), which reflect the immunological state of individuals, have been found in persons affected by asbestosis [Turner Warwick and Parker, 1970]. Research is proceeding on the detection of possible susceptibility to cancer development. The possibility of individual factors affecting susceptibility is suggested by the finding that even in groups most at risk of developing mesothelioma tumours occur in only 5 per cent: this compares with a risk of over 90 per cent developing bladder cancer in the groups most exposed to beta-naphthylamine [Williams, 1962, quoted by Gilson, 1973].

## 3. Prevention of Disease

Sufficient has been said above to indicate that the harmful effects of asbestos are severe. The first aim in prevention must be substitution by safer materials. Nowhere is this more important than in the use of crocidolite, which should be replaced by other forms of asbestos in all but essential application, and rigid control of working conditions exercised in the latter event. This is because, on the evidence so far available, diffuse mesothelioma is particularly associated with this form of asbestos, and the relevant exposure may be short (in some cases this has been a few weeks only).

With regard to lung cancer and asbestosis there is evidence of dose response, which indicates that adequate environmental control should reduce the risks to an acceptable level. These measures are most effectively applied in the factory situation, and success in reducing the incidence of lung cancer in a large asbestos factory, together with a reduction in new cases of asbestosis, has been reported by Knox, Holmes, Doll and Hill [1968]. The problems for the hygienist are greatly increased in relation to the work of laggers and insulation workers; yet here the need for effective control is emphasised by figures of new certifications between 1955 and 1963 by four Pneumoconiosis Medical Panels [McVittie, 1965]:

| | | |
|---|---|---|
| Opening, disintegrating | | 41 |
| Insulating: | Laggers | 72 |
| | Sprayers | 13 |
| | Mattress makers | 5 |
| | Others | 12 |
| Weaving | | 16 |
| Carding, spinning, etc. | | 37 |
| Slab and pipe making | | 20 |
| Brake lining | | 4 |
| Miscellaneous | | 27 |

Where dust control is unattainable protective clothing and respirators should be worn. Respirators must conform to minimum standards of dust filtration so that a false sense of security will not be engendered. Adequate provision must be made for changing of clothes and servicing of masks so that contamination is avoided, and for washing of hands. Food must be taken away from the workplace.

(In this connection the possibility of ingestion of fibres via the gastro-intestinal tract as a cause of disease needs to be considered, because of the occurrence of peritoneal mesothelioma. There is also a report by one American group of workers [Hammond, Selikoff and Churg, 1965] of an increased incidence of gastro-intestinal cancers in insulation workers studied by them, but this has not been supported by investigation in the United Kingdom [Senior Medical Inspector of Factories' Advisory Panel, 1967: Doll, 1955]. It would be expected that asbestos fibres would pass through the alimentary tract in the same way as other indigestible residues, but the very fine fibres of crocidolite might achieve penetration. The approved respirator would give adequate protection from this possibility.)

## 4. Monitoring the Health of Workers

It is self evident that the success or failure of measures to prevent asbestos induced disease will be measured by the occurrence of disease or death from these causes in future years. It may be expected that suitable examination of workers may reveal evidence of disease at an earlier stage, and lead to an earlier appraisal of the methods of prevention. It may be hoped that this will also be beneficial to the workers in two ways: by enabling any available treatment to be given at an earlier stage with an improved chance of success; and, by removing workers from the harmful environment, to prevent further deterioration in health. These propositions need to be examined.

(a) **Appraisal of control.** This can be assessed in relation to disease that is dose related, and the most sensitive index will be those aspects of disease which are capable of detection at an early stage. In the present state of knowledge this precludes examination designed to detect lung cancer or mesothelioma, which can only adequately be evaluated, in relation to asbestos exposure, by a mortality study. The early signs of asbestosis however are sufficiently characteristic to make medical examination worth while as a means of assessing control. The appearance of pleural plaques in chest radiographs has yet to be evaluated for this purpose. Factors to be considered in medical examination for appraisal of control will be discussed later.

(b) **Early institution of treatment.** No treatment is available for asbestosis. The success of treatment for lung cancer depends on other factors than early diagnosis, for example the type of cancer, squamous carcinoma being the most slow growing, and also individual factors affecting longevity in families. In a controlled trial of mass radiography at six-monthly intervals [Brett, 1969] cases found in the trial survived several months longer than a control group who presented with symptoms, but this does not take account of the time that would have elapsed before symptoms would have appeared in the cases found by mass radiography; when this is taken into account the increased survival time largely disappears. In any case chest radiography of all asbestos workers at such frequent intervals would be impracticable. Routine medical examination of asbestos workers cannot be justified on grounds of institution of treatment.

(c) Prevention of deterioration of disease. It is self evident that this cannot apply in respect of lung cancer or mesothelioma. Asbestosis is a progressive disease for which no treatment is available, and established cases do not stop deteriorating when they are removed from exposure. It may be that, as the disease normally develops insidiously over a period of several years, an early stage may be detected at which the process will be halted if exposure ceases. This will involve compensation issues for those persons advised to change their occupation, which will emphasise what is already a crucial factor in relation to control by this method, the certainty of diagnosis. It would be essential to continue examination of removed workers to assess the success of the policy, and on scientific grounds would ideally require controls, who were not so removed. This would lead to ethical problems. In spite of these difficulties arrangements of this kind have been put into action in the United Kingdom both by private companies [Lewinsohn, 1972] and on a limited basis by statutory regulation for workers in specified occupations [the Silicosis and Asbestosis (Medical Arrangements) Scheme, 1931]. The justification for health monitoring on these grounds therefore remains unproved, and merits investigation.

## 5. The Department of Employment Survey of Asbestos Workers

With the implementation of the Asbestos Regulations [1969] the Senior Medical Inspector of Factories (now Chief Employment Medical Adviser) undertook to mount a survey of asbestos workers with a view to assessing the success of control in protecting exposed workers [Annual Reports of H.M. Chief Inspector of Factories, 1970].

The desiderata for such a survey are as follows:

(a) Environmental monitoring. In a prospective study valuable information will be lost if reasonable estimates of actual asbestos dust exposure are not available for future correlation with medical data.

(b) Industrial history. The duration of exposure forms the other leg, with environmental monitoring, on which dose response relationships will be evaluated.

(c) Mortality study. It is essential that all persons, or a sufficiently large sample, should be identified at death so that diseases, such as lung cancer and mesothelioma, which are not capable of early detection and may in any case not occur until after retirement, can be related to known asbestos exposure. This aspect of study is very long term but none the less essential.

(d) Medical examination. The scope of medical examination must be so designed as to achieve the best chance of early detection of asbestosis. The following factors may be considered:

The presence of crepitations on auscultation of the chest.

The presence of finger clubbing.

Measurement of ventilatory function [$FEV_1$/FVC ratio).

Measurement of gas transfer factor.

Chest radiography.

The present of crepitations on auscultation is generally regarded as a useful sign; the sounds can be heard at the lung bases in cases with little, if any, other evidence of disease. As the disease progresses crepitations become more apparent, and increase in extent to be audible over wide areas of the chest. By this time the disease is readily detectable on the chest radiography, and the sign therefore is unsuitable for use as a quantifiable parameter.

Finger clubbing is inconstant, even in advanced cases, and is thus of no value as a criterion of early diagnosis.

Lung function tests have the advantage of introducing a quantifiable measurement. Their importance in early diagnosis must be regarded as still open to question. This is partly because of difficulties in interpretation when other factors such as chronic bronchitis, emphysema or the effects of smoking are present. The place of FEV measurement and FEV/FVC ratio in early diagnosis therefore remains to be evaluated. Assessment of gas transfer factor requires complex apparatus which limits its use to establishments which are suitably equipped. Hunt [1965] considers that reduction in gas transfer factor is the earliest sign, and Leathart [1968] has also stressed the importance of this measurement, but found that clinical signs may precede the defect.

Radiological changes were thought by Lewinsohn [1972] to be the most useful evidence for early diagnosis. He stressed the need for good quality of radiographic technique, and an exacting standard of X-ray reading. The first attempt to put the radiological interpretation of changes associated with asbestos exposure into a concise and reproducible scheme was published in 1970 [UICC/Cincinnati classification]. This has since been incorporated in the International Classification [ILO U/C International Classification of the Pneumoconioses, 1971]. Using this system it is possible that radiological diagnosis can be achieved at an earlier stage than has previously been possible. Disagreements between observers occur, especially in early changes, but considerable experience now exists in dealing with problems of inter-observer error in radiographic interpretation.

(e) <u>Smoking history</u>. The importance of smoking history, particularly in relation to lung cancer, has already been stressed. Such a history has often been difficult to obtain in retrospective studies, and it is essential that in a prospective study the smoking habits of persons included should be recorded.

The Department of Employment Study has been planned in three stages. In the first phase a pilot study of asbestos exposed persons with surveillance of ten years or more was undertaken by examining serial radiographs. This study demonstrated the feasibility of grading radiographs in accordance with the

International Classification with a view to assessing progression of the disease. The second phase is a prospective study of workers employed in selected establishments which are large users of asbestos and have their own works medical officers. Detailed occupational histories are available, and the type of asbestos to which workers are, or have been exposed, is known. The examination record includes a short health questionnaire and detailed smoking history, and the presence or absence of basal crepitations is noted. All the data thus derived are being entered on computer tape. Chest radiographs, initially screened for immediate action by works medical officers, are being classified into the extended International Classification by three-man reading panels, reading independently. Environmental tests are being carried out by the Industrial Hygiene Laboratory of H.M. Factory Inspectorate. All examinations and environmental measurements, which began in 1971-72, are to be repeated every two years, and the study is open ended. All persons included in the study are being entered into a mortality study. Arrangements have been made for additional data, in particular the results of lung function tests, to be incorporated with the basic input data, where available.

The third phase, at present in the planning stage, will extend the survey to all factories handling asbestos, and aims to cover eventually all persons who may be exposed. The first step will be to establish an asbestos register. This will be used as the basis for the arrangements for repeat examination at two-yearly intervals, and will also enable all those exposed to be entered into the mortality study immediately. The practical difficulties of covering such a large and scattered population by environmental monitoring will probably necessitate measurements being taken on a sample basis according to the type of application involved.

(f)  **Summary of medical aspects.** The serious nature of the effect of asbestos on the human body cannot be disputed. The above discussion shows that the place of routine medical examinations has yet to be evaluated. The only protection to these workers is that afforded by engineers and hygienists in controlling the environment. In this connection the special position of crocidolite demands the most rigid measures of control.

While the value of medical examination as a means of protection is in doubt the use of a medical survey may in due course reveal whether routine examination can be of direct benefit to the worker, by indicating a stage of disease at which removal from further exposure will prevent further progression of the disease occurring.

## REFERENCES

Berry, G., Newhouse, M.L. and Turok, M. [1972]. Combined effect of asbestos exposure and smoking on mortality from lung cancer in factory workers. Lancet, 2, 476-479.

Beverley, W.H.A. [1973]. A study of discernible changes in a group of asbestotics before and after certification. Journal of the Society of Occupational Medicine, 23, 61-65.

Brett, G.Z. [1969]. Earlier diagnosis and survival in lung cancer. British Medical Journal, 4, 260-262.

Buchanan, W.D. [1963]. The association of certain factors with asbestosis. XIV International Congress on Occupational Health, Madrid. Amsterdam: Excerpta Medica Foundation.

Case, R.A.M. and Hosker, M.E. [1954]. Tumour of the urinary bladder as an occupational disease in the rubber industry in England and Wales. British Journal of Preventive and Social Medicine, 8, 39-50.

Doll, R. [1955]. Mortality from lung cancer in asbestos workers. British Journal of Industrial Medicine, 12, 81-86.

Doll, R. [1971]. The age distribution of cancer: implications of models of carcinogenesis. Journal of the Royal Statistical Society, 134, 133-155.

Gilson, J.C. [1973]. Asbestos Cancer: past and future hazards. Proceedings of the Royal Society of Medicine, 66, 395-403.

Hammond, E.C., Selikoff, I.J. and Churg, J. [1965]. Neoplasia among insulation workers in the United States with special reference to intra-abdominal neoplasia. Annals of the New York Academy of Sciences, 132, 128-138.

Harington, J.S., Gilson, J.C. and Wagner, J.C. [1971]. Asbestos and mesothelioma in man. Nature, London, 232, 54-55.

Hunt, R. [1965]. Routine lung function studies on 830 employees in an asbestos processing factory. Annals of the New York Academy of Sciences, 132, 406-420.

ILO U/C International Classification of the Pneumoconioses, 1971 [1972]. Occupational Safety and Health Series No. 22 [revised]. Geneva: International Labour Office.

Kiviluoto, R. and Meurman, L. [1970]. In: Pneumoconiosis, International Conference, Johannesburg, 1969. Ed. H.A. Shapiro, pp. 190-191. Capetown: Oxford University Press.

Knox, J.F., Holmes, S., Doll, R. and Hill, I.D. [1968]. Mortality from lung cancer and other causes among workers in an asbestos textile factory. British Journal of Industrial Medicine, 25, 293-303.

Leathart, G.L. [1968]. Pulmonary function tests in asbestos workers. Transactions of the Society of Occupational Medicine, 18, 49-55.

Lewinsohn, H.C. [1972]. The medical surveillance of asbestos workers. Royal Society of Health Journal, 92, 69-77.

McVittie, J. [1965]. Asbestosis in Great Britain. Annals of the New York Academy of Sciences, 132, 128-138.

Mancuso, T.F. and El-Attar, A.A. [1967]. Mortality pattern in a cohort of asbestos workers. Journal of Occupational Medicine, 9, 147-162.

Newhouse, M.L., Berry, G., Wagner, J.C. and Turok, M. [1972]. A study of the mortality of female asbestos workers. British Journal of Industrial Medicine, 29, 134-141.

Newhouse, M.L. and Thompson, H. [1965]. Mesothelioma of pleura and peritoneum following exposure to asbestos in the London area. British Journal of Industrial Medicine, 22, 261-269.

Selikoff, I.J., Churg, J. and Hammond, E.C. [1964]. Asbestos exposure and neoplasia. Journal of the American Medical Association, 188, 22-26.

Selikoff, I.J., Hammond, E.C. and Churg, J. [1968]. Asbestos exposure, smoking and neoplasia. Journal of the American Medical Association, 204, 104-112.

Senior Medical Inspector's Advisory Panel [1967]. Problems arising from the use of asbestos. London: Her Majesty's Stationery Office.

Turner Warwick, M. and Parkes, W.R. [1970]. Circulating rheumatoid and anti-nuclear factors in asbestos workers. British Medical Journal, 3, 492-495.

UICC (International Union against Cancer) Committee [1970]. UICC/Cincinnati Classification of the radiographic appearances of pneumoconioses. Chest, 58, 57-67.

Wagner, J.C., Gilson, J.C., Berry, G. and Timbrell, V. [1971]. Epidemiology of asbestos cancers. British Medical Bulletin, 27, 71-76.

Wagner, J.C., Sleggs, C.A. and Marchand, P. [1960]. Diffuse pleural mesothelioma and asbestos exposure in the North Western Cape Province. British Journal of Industrial Medicine, 17, 260-271.

## PATHOLOGY OF ASBESTOS

by

Professor J. Champeix,
Professor of Forensic Medicine and Occupational Health
Faculty of Medicine, Clermont-Ferrand [France]

### 1. Introduction

Until recent years it was thought that workers exposed to the inhalation of asbestos fibres were liable to be affected by a form of pneumoconiosis: asbestosis, which was defined as "broncho-pulmonary fibrosis". This form of pneumoconiosis, like other pneumoconioses causing right cardiac insufficiency, could be complicated by bouts of acute breathlessness. In most countries it was recognised that bronchial cancer occurred with significant frequency in subjects suffering from asbestosis; these countries consequently considered broncho-pulmonary cancer to be a complication of asbestosis, giving entitlement to compensation as an occupational disease.

In the last few years the problem of the pathology of asbestosis has taken on a new dimension in view of the frequency with which pleural or pleuro-peritoneal mesotheliomas have been discovered in subjects exposed to asbestos fibres.

### 2. Criteria for Detecting Asbestosis

Early detection of asbestosis as such - that is to say broncho-pulmonary fibrosis resulting from the inhaling of asbestos fibres, is difficult. The detection must be based on a set of factors, none of which by itself is significant for the purposes of diagnosis.

The factors of diagnosis are as follows:

2.1. Knowledge of a definite occupational hazard. It is necessary to analyse:

- the duration of exposure;

- the extent of the exposure by measuring the degree of atmospheric pollution;

- the type of materials used (chrysotile, crocidolite, amosite, etc.);

- the condition of work: employment in a textile factory, the use of wet materials (asbestos-cement), spraying (building), demolition work (shipyards, foundry furnaces, etc.).

2.2. The existence of subjective functional disorders determined through careful questioning: coughing, dyspnea caused by exertion, etc.

2.3. The physical signs revealed by auscultation: slight localised crepitation, first of all at the bases, revealing the existence of a certain degree of fibrosis, sometimes pleural friction.

2.4. Radiographic evidence: only standard X-ray films should be used for detection since mass miniature radio-photography is utterly inadequate for this purpose.

Radiographs of asbestos workers may show:

- fibrosis,
- pleural calcification.

(a) **Fibrosis**

The essential characteristic of fibrosis as found in asbestosis is that it is a linear fibrosis as opposed to the nodular type fibrosis found in silicosis.

In the early stages a lessened transparency can be seen in the lower parts of the lung fields caused by a slight exaggeration of the linear markings, with barely visible fine granulation, giving a "spider's web" or "misty" appearance. This appearance borders on the normal in an area where the lung outline is already highly complex: it is obviously advisable therefore for X-rays to be taken at six monthly or yearly intervals, according to the same techniques, of subjects exposed to this occupational hazard.

At the more advanced stage, the lower two-thirds of the two lung fields present a marked appearance of fibrosis, in contrast to the clearer upper parts. These appearances are constituted by an exaggeration of the linear markings, with small, uniform shadows whose outlines are ill-defined. This is the characteristic "ground glass" appearance. These pictures never have the clearness of the micro-nodules seen in silicosis. They are finer, more numerous and bear a closer resemblance to the cardiac lung. Moreover the cardiac shadow at this stage is much larger in volume: linear shadows radiate from it into the lung fields, the whole thing giving what is known as the "porcupine heart" appearance.

Between the early and the terminal pictures, all intermediary stages are visible.

(b) **Pleural calcification**

Pleural calcification occurs frequently in asbestosis, more particularly it seems in subjects who have been exposed to comparatively little dust: lagging workers, workers in textile factories where not much dust is given off, etc.

Site: generally the calcification is bilateral and relatively symmetrical. It is found in:

- the two diaphragmatic pleura;

- the latero-thoracic and pericardial parietal pleura;
- the inter-lobar pleura is nearly always unimpaired; nevertheless the lower right scissure is frequently visible on the radiograph but without a really thickened appearance.

Morphology: often linear at the level of the diaphragmatic and pericardial pleura, or rounded; in extended clusters in the latero-thoracic parietal pleura.

The calcification never resembles "cuttle-bone" calcification in form or area. It may or may not have the appearance of asbestosis-type fibrosis.

Pleural plaques are fairly frequently discovered in subjects little exposed to asbestos fibres.

In the present state of our knowledge, it would seem:
- that considerable exposure to dust causes a fibrosis;
- that relatively little exposure may cause the appearance of pleural calcification.

2.5. The presence of asbestos bodies in sputum.

Morphology: The characteristic colour is normally golden yellow, brownish yellow or greenish.

The size varies but is generally between 10 and 15 µm. It is not unusual, however, to see larger bodies: 50-100 µm. They range from 3 to 12 µm in thickness.

Their shape is very unusual. Generally they are rectilinear, dumb-bell shaped at the ends, though they may also be club-shaped at just one end or nail-shaped with several swellings, hence the moniliform, curved or angular appearance with several swellings.

They also have their own special structure. The unit formed by the asbestos body is made up of a central axis, a colourless asbestos needle, which is almost completely surrounded by an endogenous capsule. It is this capsule, newly formed around the asbestos needle, that gives these bodies their yellow colour. The thicker the capsule the deeper the colour.

Little is known of the nature of this capsule. In all probability it is proteino-ferruginous. The iron would explain the golden yellow colouring; most of this iron is thought to be of endogenous origin (an analogy may be made here between the colouring of these bodies and the ocre pigmentation of the spleen). The iron might come, as is the view of Sundius and Bydgen, from the haemoglobin decomposition in the fibre.

In explaining the differences in the distribution of this capsule over the fibre, reference must be made to the role of the breaks, the "breakdown of the crystalline network of the asbestos fibre". At the breakage point there is thought to be a considerable difference in potential that would encourage certain micro-reactions. The breakage is more marked at the extremities, which explains why the proteino-ferruginous substances have an affinity for the extremities of the fibres. Some breakages, which are less considerable along the length of the fibre, would explain some of the moniliform appearances.

Moreover, by creating a sufficient difference in potential, an asbestos body can be artificially produced from asbestos fibre in the presence of proteins.

What must be remembered in diagnosis is that sputum tests must be repeated. A negative test does not necessarily imply the absence of asbestos bodies.

- Although subjects exposed to asbestos dust present symptoms of coughing, expectoration is unusual. In order to detect the presence of asbestos bodies, tests should be carried out when there is a slight expectoration such as when there is bronchial catarrh during influenza.

- Asbestos bodies appear early after repeated exposure to asbestos dust; they are found in subjects who have been exposed for several months.

- When a few isolated bodies (no more than two or three) are found, this merely signifies that the subject has inhaled asbestos dust and does not imply a definite pathological condition. Systematic sputum tests in asbestos workers reveal such isolated bodies in nearly all subjects even if there is no radiological or functional evidence of asbestosis.

- Grouped together, they seem more significant; groups of 20 to 30 bodies may be observed in a radial position, the bulbous extremity pointing outwards.

2.6. Respiratory function tests.

The functional syndrome observed during asbestosis is often characteristic and related to the presence of diffuse interstitial fibrosis. As a result of the research work carried out by Cournand, Baldwin, Austrian, Riley, etc., it is known that diffuse fibrosis entails impairment of oxygen diffusion. Carbon monoxide methods of studying gas diffusion have enabled this impairment of alveolar-capillary diffusion to be confirmed.

- Spirographic data. In simple asbestosis, disorders of a restrictive type are observed: there is a homogenous drop in all volumes, mostly of the VC and TC. The residual volume is not increased, but, on the contrary, may be lowered proportionately to the VC. The total capacity (vital capacity + residual volume) is low.

The FEV is lowered at the same time as the VC and the ratio of the forced expiratory volume to the vital capacity is normal. Sometimes, however, bronchial symptoms are associated with simple asbestosis and an obstructive syndrome may be associated with the restrictive syndrome.

- Ventilatory data. When the subject is at rest hyperventilation is observed. This is due to a high respiratory frequency rather than to an increase in the current volume. Oxygen consumption remains lower than what it should be and the respiratory equivalent is high. Frequently there is alveolar hyper ventilation but ventilatory efficiency (the alveolar ventilation/total ventilation ratio) is lowered.

- Gas exchange. Gas exchange disturbance observed in asbestosis, as in other diffuse interstitial pulmonary fibroses, affects the alveolar-capillary block. Characteristic symptoms are hypoxaemia and hypercapnia, associated with the simple restrictive syndrome. Exercise tests are required to reveal the presence of hypoxaemia, which is not present during rest, since the $PaCO_2$ remains normal or low. These disturbances are due to the thickening of the alveolar-capillary septum or, according to Rossier, to the reduction of the pulmonary vascular bed which lessens the time of air-blood contact. The thickening of the membrane impedes oxygen diffusion whereas $CO_2$, which is much more diffusable, passes easily.

The study of the carbon monoxide diffusing capacity is of interest. Sometimes the carbon monoxide diffusing capacity is considerably lowered. Diffusion disturbance can also be assessed by studying the $PO_2$ alveolar-arterial gradients. The $PO_2$ gradient in hypoxia is of interest since it enables the oxygen diffusing capacity to be calculated. The oxygen diffusing capacity measurement method is not widely used because of its complexity and the equipment required, but it provides valuable information on disturbances in gas exchange.

In practice, in cases of pure asbestosis, where the extent of diffuse fibrosis lesions could be ascertained anatomically, functional tests carried out in the course of the patients' lifetime gave results that did not fully correspond to this pattern.

Nevertheless this respiratory functional investigation is extremely worth while since it can be carried out before lesions become apparent radiologically.

- The lung scintigraphy. In association with the techniques described above, this is useful in studying the degree of lung perfusion. Frequently it is found that the peripheral areas are inadequately perfused.

### 3. Problems Presented by Exposure to Asbestos Fibres and the Appearance of a Pleural Mesothelioma

At the moment this is the problem that raises the most questions, which is why it is the main issue dealt with in this paper.

3.1. Statistics

3.1.1. How have the statistical studies been carried out?

The statistical study of a relationship between a mesothelioma and exposure to asbestos cannot be accurate because:

(a) there is always an element of doubt concerning the diagnosis itself;

(b) in some cases it is difficult to prove that there has really been exposure to asbestos; this may be because the records of certain firms have been destroyed, because of a high degree of labour mobility in some countries, or yet again because of the ignorance of the subject himself or of his family or even because of incompatability between the statistician and his subject;

(c) there is generally an interval - sometimes lasting several decades - between exposure and sickness, so that the subject may die of something else or not yet be affected when the study is carried out.

There are two ways of conducting a study:

(1) Upstream: an investigation is made among patients suffering from mesotheliomas to determine whether they have ever been exposed to asbestos dust in the following contexts: occupational, by questioning undertakings, the patient himself or his family and friends; residential, in the fairly close neighbourhood of a source of pollution such as a mine or a processing plant; domestic, when a member of the household comes home every day covered with asbestos dust.

(2) Downstream: a record is kept of cases of malignant mesotheliomas among an exposed population group.

A - Various types of exposure to asbestos:

(a) occupational exposure:

- asbestos extraction [mines]

- ore working [pulverizing, sorting ...]

- ore transportation [dockers, sailors ...]

- processing, manufacture of the ore [spinning, manufacture of numerous products by mixing asbestos with another refractory cement body, brakes, linoleum, electrical and thermal insulation, fireproof materials ...]

- use of these various products:

    insulation - in building
              - in shipyards
              - in explosives-manufacturing
                works during the last war;

    various unusual cases such as road-making with waste materials containing traces of asbestos.

(b) **other forms of exposure**:

- neighbourhood, when the past or present dwelling is near to a source of pollution,
- domestic,
- through atmospheric pollution, which is open to question, disputable and problematic.

(c) **multiple exposure**:

In some areas, where the asbestos industry is highly diversified and there is rapid labour turnover, as in Dresden (19) or Belfast (13), most of the population are exposed directly or indirectly.

(d) **uncertain exposure**:

This is the case of a patient in respect of whom exposure has not been proved although asbestos has been found in the form of asbestos bodies, either in the sputum (in order for the bodies to be present in the sputum there must be massive exposure which, therefore, is generally known), or in the lungs (during biopsy or autopsy). Even then it is necessary to ascertain that these really are asbestos bodies and here too misinterpretation is possible. This problem will be studied below.

(3) **In the course of both these types of study** other possible causes of cancer in the patients must also be sought and the following points should be investigated:

(a) any history of neoplasia in the family,

(b) the patient's previous history as regards:

- asbestosis lesions
- pleurisy
- collapse therapy for the treatment of tuberculosis
- exposure associated with other products (mining, ore processing, explosives, transport).

(c) the patient's smoking habits, which deserve special consideration.

(d) the patient's home environment with the problem of atmospheric pollution by products other than asbestos.

3.1.2. **Have mesothelioma patients been exposed to asbestos**?

A - **Method of study**

(1) Determination of patients affected:

- Newhouse makes investigations in pathology departments where cases have been kept for several decades [23].

- McDonald writes to members of the Canadian association of pathologists to know whether any fatal cases of pleural or peritoneal tumours have been reported. In the affirmative, an inquiry is carried out to isolate cases of real mesothelioma [12].

- Others investigate cases reported in hospitals [14].

(2) Verificiation of the diagnosis, which, as has been seen is a thorny problem.

(3) Investigation of exposure to asbestos by inquiries that are long, difficult and sometimes impossible.

The inquiries must be made: in the undertaking in which the patient has worked, among the patient's family and friends, of the patient himself wherever possible, and asbestos bodies must be sought in the lung, which sometimes gives the only indication of exposure [22].

(4) A population group not affected by mesothelioma must be studied in order to determine whether there has been any exposure to asbestos at any point during their lives.

(5) If possible conclusions must be drawn from this study.

In order to have an idea of the difficulties encountered, one may consider the work of Dalquen [9] who, after studying 119 cases, obtained:

- in the form of occupational information:

    full information in respect of 66 cases, incomplete information in respect of 32 and no information in respect of 21;

- in the form of information on the place of residence, including various changes of residence that may have taken place:

    full information in respect of 26 cases and incomplete information in respect of 83.

B - **The Figures**

1. Over-all study

TABLE No. 1

| Source of figures | Number of cases | Exposure to asbestos | | | |
|---|---|---|---|---|---|
| | | Definite | Accidental or probable | Possible | None or improbable |
| [1]  | 22  | [..........20......]       |     |    | 2  |
| [6]  | 15  | 2   |     |    | 13 |
| [9]  | 119 | 17  | 38  |    | 43 |
| [10] | 5   |     |     |    | 5  |
| [13] | 42  | 32  |     |    | 10 |
| [14] | 73  | 20  | 20  | 27 | 5  |
| [15] | 23  |     |     |    | 23 |
| [12] | 93  | 10  | 9   | 17 | 57 |
| [21] | 11  | 7   |     |    |    |
| [30] | 76  | 40  |     |    | 36 |
| [25] | 1   | 1   |     |    |    |
| [26] | 1   | 1   |     |    |    |
| [27] | 76  | 3   | 3   |    |    |
| [28] | 22  | 18  |     |    |    |
| [29] | 148 | [..........124..................] | | | 24 |
| [30] | 40  | 35  |     |    | 5  |
| [30] | 12  | 8   | 2   |    |    |
| [22] | 15  | 11  | 3   |    | 1  |
| TOTAL | 794 | 205 | 219 | 44 | 224 |
| + | 100 | 25.6 | 27.3 | 5.5 | 28.0 |

Determination of exposure to asbestos corresponds roughly to that of McEwen [14].

- definite: direct information on work with asbestos.

- probable: the information obtained gives reason to believe that the subject has been in contact with asbestos.

- possible: the information gives reason to believe that the subject has had dealings with people who have been in contact with asbestos.

- none: exposure appears highly unlikely.

This classification, however, is not followed by all and various alternatives are possible:

- The sum of the figures relating to asbestos exposure is lower than the total number of cases: this is due to lack of information.

- In one case out of four there appears to have been definite exposure to asbestos, while in more than one case out of two there is a possibility of exposure and in one case out of four there is no exposure.

These figures were compared with a control series, that is to say a population group, not affected by mesothelioma, in which the sexes and age groups are in the same proportion. The study was carried out in the same way.

TABLE No. 2:
EXPOSURE OF PATIENTS NOT SUFFERING FROM MESOTHELIOMA

| Source | Diseases affecting the control group | Number of cases | Definite | Accidental or probable | Possible | None or improbable |
|---|---|---|---|---|---|---|
| [12] | Secondary lung tumour | 88 | 1 | 2 | 19 | 66 |
| [12] | Primary lung tumour | 88 | 1 | 3 | 14 | 70 |
| [14] | Cancer | 73 | 0 | 12 | 34 | 27 |
| [14] | Cardio-vascular | 72 | 3 | 5 | 18 | 46 |
| | TOTAL | 321 | 5 | 22 | 85 | 209 |
| | PERCENTAGE | 100 | 1.6 | 6.9 | 26.4 | 65.1 |

A comparison of tables nos. 1 and 2 clearly reveals that a population group affected by mesothelioma has had much greater exposure to asbestos than a population group affected by any other disease (the total of cases of definite to possible exposure comes to 58.4 per cent, as against 34.9 per cent).

### 3.1.3. Detection of mesothelioma in a population group exposed to asbestos

This too entails a long and difficult study since healthy people must be followed for several years because, as will be seen, mesothelioma takes a very long time to declare itself.

In undertakings where the hazard is known to exist, industrial physicians keep records of the exposed workers and try to keep trace of them even if they leave for other jobs.

Another method is also possible: The detection of mesothelioma in an area where work involving asbestos is widespread, as in South Africa or in certain ports in Great Britain.

TABLE No. 3

| Source | Background | Population | Mesothelioma | Percentage |
|---|---|---|---|---|
| [13] | 30 deaths among asbestos workers in Belfast | 30 | 2 | 7 |
| [18] | 380 deaths in a New York undertaking from January 1943 to December 1968 | 380 | 22 | 0.58 |
| [19] | 150 deaths among exposed workers in Dresden, 132 of whom were known to have asbestosis | 150 | 6 | 4 |
| [20] | Antophyllite mine in Finland: this form of asbestos does not seem to be dangerous | 1 000 | 0 | 0 |
| [11] | McDonald in Canada, in a mining region: 11 788 persons followed, among whom there were 2 457 deaths | 2 457 | 3 | 0.123 |
|  | In a subsequent study the percentage is slightly higher: 771 more deaths, i.e. an average percentage of 0.155 | 771 | 2 | 0.26 |
| [27] | Population group in the North of Italy [chrysotile mine and processing plant] |  |  |  |
|  | - 172 deaths among asbestosis patients | 172 | 3 | 1.8 |
|  | - 544 asbestosis patients living | 544 | 2 | 0.37 |
|  | In this latter group the number of cases of mesothelioma may still increase |  |  |  |
|  | TOTAL | 3 504 | 40 | 1.14 |
| TOTAL excluding [20], this being a special type of asbestos |  | 2 504 | 40 | 1.6 |

Comments:

If one considers workers who entered the factory after 31 December 1942, there are no cases of mesothelioma. Is this because of better protection, because mesothelioma had not yet declared itself or a matter of chance?

The mesotheliomas are pleural, peritoneal ....

Percentages vary from one study to another but 1.5 per cent is a figure that may be borne in mind.

Is_this_figure_significant?

Vigliani [27] made a study in a highly industrialised region of Northern Italy of the average frequency of mesothelioma. He performed 24 700 autopsies and obtained an average percentage of 0.30.

Dalquen [9] studied mesothelioma in relation to environment: out of a population of 1 857 371 in the Hamburg region, he obtained a proportion of 0.56 per cent.

If one compares either 0.30 per cent or 0.56 per cent with 1.5 per cent, asbestos definitely appears to be involved.

The question remains as to whether asbestos alone is responsible or whether it forms part of a set of conducive factors.

3.1.4. Conditions_of_occurrence_of_mesotheliomas

A - The_type_of_exposure, wherever it has been possible to determine this:

## TABLE No. 4: HISTORY OF EXPOSURE TO ASBESTOS OF PATIENTS WITH MESOTHELIOMAS

| Source | Number of cases | Industry | Mining | Insulation | Handling | Domestic | Other types of contact | Environment |
|---|---|---|---|---|---|---|---|---|
| [1] | 22 | 13 | - | - | - | - | 1 | - |
| [6] | 2 | 1 | - | 1 | - | - | - | - |
| [9] | 119 | 29 | - | 5 | 17 | - | 4 | - |
| [14] | 80 | 50 | - | 1 | 9 | - | - | 6 |
| [16] | 240 | 15 | - | 53 | 48 | 11 | 98 | 15 |
| [12] | 30 | 4 | 3 | 5 | 1 | 11 | 3 | - |
| [23] | 76 | 24 | - | 7 | - | 9 | - | - |
| [25] | 1 | 1 | - | - | - | - | - | - |
| [30] | 52 | 36 | - | - | 3 | - | 6 | - |
| TOTAL | 622 | 173 | 3 | 72 | 75 | 31 | 112 | 21 |
| TOTAL CASES OF KNOWN EXPOSURE: | 487 | | | | | | | |
| % | 100 | 35.5 | 0.6 | 14.8 | 15.4 | 6.4 | 23.0 | 4.3 |

There may have been multiple exposure. Only the principal types of exposure have been taken into consideration in this table (the main sources being given in the first columns).

It should also be noted that the heading "Insulation" includes workers involved in the manufacture and use of insulating materials as well as those who merely come into contact with them.

The "Other types of contact" are numerous and unusual including, for example, women who have repaired sacks that have contained asbestos [30].

Here too, these figures should be compared with those of a control series, an aspect about which little has been written.

TABLE No. 5: EXPOSURE TO ASBESTOS IN A
CONTROL SERIES NOT AFFECTED BY MESOTHELIOMA

| Source | Criterion in selecting the control series | Number of cases | Industry | Mining | Insulation | Handling | Domestic | Other Contacts | Environment |
|---|---|---|---|---|---|---|---|---|---|
| [1] |  | 46 | 19 |  |  |  |  |  |  |
| [14] | Carcinoma | 71 | 51 |  | 0 | 2 |  |  | 37 |
|  | Cardio-vascular | 72 | 48 |  | 0 | 5 |  |  | 35 |
| [34] | Primary lung tumour | 10 | 0 | 1 | 0 | 0 | 7 | 2 |  |
|  | Secondary lung tumour | 13 | 0 | 1 | 0 | 0 | 9 | 3 |  |
| [23] |  | 76 | 2 |  | 4 | 2 |  | 1 |  |
|  | TOTAL | 288 | 120 | 2 | 4 | 9 | 16 | 6 | 72 |
|  | PERCENTAGE | 100 | 41.6 | 0.7 | 1.4 | 3.1 | 5.6 | 2.1 | 25 |

Comparison of tables Nos. 4 and 5 reveals that although the risk is high in industry, this is solely because a large number of workers are exposed there.

On the other hand in insulation and handling, where there is perhaps greater and more direct contact, there seems to be greater exposure to the risk of mesothelioma.

McDonald [31] has reached the conclusion that exposure in industry is far more dangerous than in extraction or crushing processes. But Canadian asbestos is a variety of chrysotile, which does not seem to be dangerous, and in industry it is mixed with crocidolite.

Consequently the type of exposure alone does not provide an adequate explanation.

B - Degree of Exposure

TABLE No. 6

| SOURCE | NUMBER OF CASES | EXPOSURE Large-scale | Moderate | Slight | None or unknown |
|---|---|---|---|---|---|
| [13] | 42 | 8 | | 24 | 10 |
| [25] | 1 | 1 | | | |
| [22] | 15 | 8 | 1 | 2 | 4 |
| TOTAL | 58 | 17 | 1 | 26 | 14 |
| PERCENTAGE | 100 | 29.3 | 1.7 | 44.8 | 24.1 |

Slight exposure is therefore just as dangerous, if not more so than large-scale exposure. Webster [29] reached the same conclusion by calculating the percentage of deaths from mesothelioma in relation to the degree of exposure.

TABLE No. 7

PERCENTAGE OF DEATHS IN RELATION TO
THE DEGREE OF EXPOSURE

| EXPOSURE PER CASE OF MESOTHELIOMA | None | Slight | Moderate | Marked |
|---|---|---|---|---|
| | 1.0 | 2.8 | 3.6 | 2.9 |

C - Length of Exposure

TABLE No. 8

|  | Length of Exposure (Years) | | | |
| --- | --- | --- | --- | --- |
|  | Minimum | Average | Maximum | Real |
| [1]  | 6 months |  | 30 |  |
| [6]  |  |  |  | 2 |
| [9]  | 14 days | 12.2 | 41.0 | 35 |
| [17] |  |  |  | 1 |
| [12] | 16 |  | 50 |  |
| [25] |  |  |  | 35 |
| [26] |  | > 20 |  |  |
| [27] |  |  |  | 21 |
| [30] | 3 | 16 | 50 |  |

It would seem that exposure may sometimes be of very short duration, or even a matter of chance (14 days in one case) though the average is around 20 years.

In general it appears that exposure over several decades is necessary for, in the crocidolite mines in Australia, where the labour force changes frequently and is consequently exposed for no more than an average of two to three years, very few cases of mesothelioma have been reported among thousands exposed.

In addition according to Selikoff [18], the degree of exposure would seem to affect the site of the mesothelioma since short-term exposure would appear to affect the pleura, whereas long exposure appears to cause peritoneal neoplasia.

## D - Time lapse between exposure and discovery of a tumour

### TABLE No. 9:
### TIME LAPSE BETWEEN THE BEGINNING OF EXPOSURE AND DISCOVERY OF THE TUMOUR

| Source | Number of Cases | Groups | Beginning of Exposure - Tumour (Years) | | |
|---|---|---|---|---|---|
| | | | Minimum | Average | Maximum |
| [9] | 16 | Asbestos workers | 11.0 | 30.8 | 42.0 |
| | 14 | Sailors and shipbuilders | 25.0 | 40.6 | 48.0 |
| [18] | 6 | Pleural mesotheliomas | 25 | 34.8 | 38 |
| | 16 | Peritoneal mesotheliomas | 32 | 43.0 | 61 |
| [27] | 6 | | 16 | | 40 |
| [30] | 33 | | 13 | 42 ± 12 | |

Average: 40 ± 15 years

### TABLE No. 10:
### TIME LAPSE BETWEEN THE END OF EXPOSURE AND DISCOVERY OF THE TUMOUR

| | Average | | Actual time | | | |
|---|---|---|---|---|---|---|
| Source | [9] | [30] | [6] | [17] | [25] | [26] |
| Number of years | 35.2 | 20 years | -0 -24 | 25 | 0.5 | 5 |

It is of interest to note that a mesothelioma may appear several decades after exposure has ceased.

E - Sex of mesothelioma patients (exposed to asbestos or not)

TABLE No. 11

| SOURCE | NUMBER OF CASES | MALE | FEMALE |
|---|---|---|---|
| [6]  | 15  | 11  | 4  |
| [9]  | 119 | 79  | 40 |
| [10] | 5   | 4   | 1  |
| [14] | 80  | 73  | 7  |
| [16] | 390 | 309 | 81 |
| [17] | 1   | 1   | 0  |
| [15] | 23  | 16  | 7  |
| [19] | 6   | 0   | 6  |
| [23] | 56  | 31  | 25 |
| [24] | 39  | 31  | 8  |
| [25] | 1   | 1   | 0  |
| [26] | 1   | 1   | 0  |
| [27] | 6   | 4   | 2  |
| [29] | 175 | 130 | 45 |
| [30] | 56  | 34  | 22 |
| TOTAL | 973 | 725 | 248 |
| PERCENTAGE | 100 | 74.4 | 25.6 |

In three cases out of four the patients are men, whether in exposed or non-exposed population groups.

Is this because men are more exposed, because they are more vulnerable or because they are exposed to other conducive toxins, particulary tobacco?

F - Association of exposure to other products

(a) Smoking habits

We shall now consider the possible consequences on the occurrence of a mesothelioma of the number of cigarettes smoked in 24 hours by a smoker exposed to asbestos:

TABLE No. 12:
SMOKING HABITS OF MESOTHELIOMA PATIENTS
WHO HAVE BEEN EXPOSED TO ASBESTOS

| SOURCE | NUMBER OF PACKETS | | | |
|---|---|---|---|---|
| | 0 | $\simeq 1/2$ | $\simeq 1$ | >1 |
| [14] | 12 | 30 | 18 | 12 |
| [12] | 2 | 3 | 13 | 1 |
| [26] | 0 | 1 | 0 | 0 |
| [30] | 10 | 8 | 10 | 10 |
| TOTAL | 24 | 42 | 32 | 23 |
| PERCENTAGE | 21 | 34 | 25 | 20 |

TABLE No. 13: SMOKING HABITS OF MESOTHELIOMA
PATIENTS WHO HAVE NOT BEEN EXPOSED TO ASBESTOS

| SOURCE | NUMBER OF PACKETS | | | |
|---|---|---|---|---|
| | 0 | $\simeq 1/2$ | $\simeq 1$ | >1 |
| [12] | | | | |
| Men | 17 | 17 | 35 | 19 |
| Women | 27 | 7 | 10 | 1 |
| TOTAL | 45 | 34 | 45 | 20 |
| PERCENTAGE | 31 | 25 | 31 | 13 |

## TABLE No. 14:
### SMOKING HABITS OF A CONTROL SERIES UNAFFECTED BY MESOTHELIOMA

| SOURCE | NUMBER OF PACKETS | | | |
|---|---|---|---|---|
| | 0 | ≃ 1/2 | ≃ 1 | > 1 |
| [12] | | | | |
| Men   A | 19 | 19 | 34 | 15 |
| B | 4 | 10 | 43 | 30 |
| Women A | 33 | 4 | 6 | 3 |
| B | 20 | 8 | 13 | 4 |
| [37] | | | | |
| Lung tumour | 4 | 16 | 25 | 25 |
| Cardio-vascular | 4 | 21 | 22 | 21 |
| TOTAL | 84 | 78 | 143 | 98 |
| PERCENTAGE | 20.8 | 19.4 | 35.5 | 24.3 |

These results can be compared with those of Hammond and Selikoff [32] who, out of 26 cases of pleural mesothelioma found 18 smokers and 7 patients whose smoking habits were not known. No useful conclusions can be drawn from so few cases, nor from the 51 cases of peritoneal mesothelioma, consisting of:

- 9 non-smokers out of 2 066 non smokers, i.e. 0.43 per cent
- 29 smokers out of 9 590 smokers, i.e. 0.30 per cent

and 13 in respect of whom the relevant information is not available.

The only conclusion that could be drawn here would be that smoking affords protection against this disease (0.43 per cent as against 0.30 per cent)!

Thus although a smoker working with asbestos is, statistically, eight times more likely to have lung cancer, it does not seem that his smoking habits make him any more likely to have a mesothelioma than his non smoking workmates.

The same conclusions have been reached by Whitwell [30] who found no significant difference between smokers and non smokers as regards either the degree or the duration of exposure to asbestos.

As regards Selikoff [18], all his mesothelioma cases are smokers, with the exception of four with peritoneal and not pleural diseases: no firm conclusions can be drawn from the 22 cases he studied.

(b) Exposure to other products

The ore is never completely pure. Consequently in a study carried out by Webster [29], there were 54 cases of mesothelioma involving patients who had been exposed to asbestos alone, but also 21 cases where there had been exposure to a mixture of asbestos and manganese. This seems to be important in view of the fact that two cases of pleural mesothelioma were found among manganese miners who had never been in contact with asbestos.

Furthermore Webster noticed that the percentage of mesotheliomas is considerably higher in Cape Province (South Africa), where crocidolite is extracted (as will be seen, this is a particularly dangerous form of asbestos), and where, in addition, the workers are exposed to manganese and iron.

It certainly appears that exposure to iron is significant for there is a high rate of mesotheliomas in shipyards (Stumphius' thesis) where ships were insulated with asbestos-based materials and where welders and pipe fitters were exposed, particularly during repair work, to asbestos dust combined with iron oxide gases let off by blow pipes and during arc welding.

This also raises the problem of asbestos bodies and pseudo-asbestos bodies, which problem will be reverted to later.

G - Age of subjects affected by mesothelioma

TABLE No. 15: AGE AT THE TIME OF DIAGNOSIS

| SOURCE | Minimum | Average | Maximum |
|--------|---------|---------|---------|
| [6]    | 43      | 55      | 76      |
| [6]    | 40      | 55      | 73      |
| [14]   | 28      |         | 87      |
| [17]   |         | 44      |         |
| [21]   | 34      |         | 74      |
| [24]   | 23      | M 56    | 76      |
|        |         | F 43    |         |
| [26]   |         | 65      |         |
| [30]   | 16      | 60 ± 10 |         |
| RESUME | 16      | 55      | 87      |

The difference between age and the time of diagnosis and that at the time of death is negligible since mesotheliomas generally develop very rapidly, one to two years at the most.

For McDonald [12], there is no significant difference according to age or sex as regards the incidence of mesothelioma in comparison to a control group with primary or secondary lung cancer. Moreover, in the age group below 50 neither males nor females predominate, though above 50 males predominate (table 16).

TABLE No. 16:
AGE AT DEATH

| Source | \ Age Groups | | | | | |
|---|---|---|---|---|---|---|
| | 0 | 40 | 50 | 60 | 70 | 100 |
| [9] | 1 | 4 | 15 | 37 | 22 | |
| [15] | 4 | 2 | 3 | 5 | 2 | |
| [12] Mesotheliomas: | | | | | | |
| - pleural | 9 | 9 | 18 | 26 | 18 | MEN |
| - peritoneal | 1 | 4 | 11 | 2 | 8 | |
| - pl. + perit. | | | 1 | | | |
| - pericard. | | | 1 | | | |
| [14] | 5 | 7 | 23 | 25 | 13 | |
| TOTAL M. | 20 | 26 | 72 | 95 | 61 | |
| PERCENTAGE | 7.3 | 9.4 | 26.4 | 34.7 | 22.3 | |
| [9] | 0 | 4 | 15 | 15 | 7 | |
| [14] | | 2 | 2 | 3 | 2 | |
| [15] | 1 | 2 | 4 | | | |
| [12] Mesotheliomas: | | | | | | WOMEN |
| - pleural | 9 | 7 | 13 | 5 | 9 | |
| - peritoneal | 2 | 1 | 4 | 5 | 7 | |
| - pl. + perit. | | | | 2 | | |
| - pericard. | 1 | 1 | 1 | | | |
| TOTAL W. | 12 | 14 | 27 | 34 | 25 | |
| PERCENTAGE | 10.5 | 12.3 | 24.0 | 30.2 | 23.0 | |
| [29] | 29 | 45 | 40 | 0 | 8 | MEN AND WOMEN |
| [30] | 1 | 9 | 15 | 22 | 9 | |
| M. + W. | 32 | 40 | 99 | 129 | 86 | |
| TOTAL | 62 | 94 | 154 | 151 | 103 | |
| PERCENTAGE | 11.0 | 16.5 | 27.4 | 26.8 | 18.4 | |

## H. Race

Webster, in South Africa, broke down 174 cases of mesothelioma as follows:

- Whites (pure European descent) ..................... 78
- Non-Whites (Bantu descent) ........................ 55
- Coloureds ......................................... 44

This shows a distinct predominance among the whites, which may be due to the fact that medical supervision in their case is much closer, especially in South Africa.

## I. Atmospheric pollution in non-industrial environments

There is also atmospheric pollution in big cities with a wide range of industries; any asbestos contributing to this pollution is in the form of waste, which is difficult to check (dust from car brake linings, insulation, etc.).

There is also the pollution caused by factories handling asbestos, in cases where dust is allowed to escape through inadequate preventive measures.

(a) Vicinity of factories handling asbestos.

This is difficult to assess, since exposure is from more than one source - very often the inhabitants of the districts concerned also work in the factories.

TABLE No. 17

| SOURCE | DISTANCE OF HOME FROM ASBESTOS FACTORY | | |
|---|---|---|---|
| | less than 1/2 mile | over 1/2 mile | |
| [14] | 51 | 22 | Meso-thelioma |
| [23] | 11 | 25 | |
| TOTAL | 62 | 57 | |
| PERCENTAGE | 52.1 | 47.9 | |
| [14] | 37 | 34 | Control |
| | 35 | 37 | |
| [23] | 5 | 62 | |
| TOTAL | 77 | 113 | |
| PERCENTAGE | 36.6 | 63.4 | |

It would appear that the risk of mesothelioma is greater in the immediate vicinity of an asbestos factory.

(b) Pollution in the cities.

(a) The presence of asbestos in the air of cities has not been widely investigated, but it is detected through the discovery of asbestosic bodies in the lungs of city-dwellers: 20 per cent in the case of the male population of Belfast [13], whereas direct exposure is insignificant. This problem will be reverted to in connection with asbestosic bodies.

(b) The frequency with which mesotheliomas are encountered outside sections of the population which are not particularly exposed is subject to striking variations:

In Northern Italy, whereas the average is around 0.30 per cent [27], Vigliani, after 24 700 systematic autopsies, found a frequency of:

- 0.45 per cent in Turin
- 0.24 per cent in Milan
- 0.05 per cent in Pavia

(whereas Turin and Milan are highly industrialised, Pavia is a small town in a farming area),

Dalquen [9] made a similar investigation in the Hamburg area of Germany (see table 18).

A quite recent study by Bohling [3], also in the Hamburg area, shows, by mapping the cases of mesothelioma not due to occupational exposure, that there is a marked concentration around the sources of pollution, with an appreciably higher concentration among the population down wind.

In the vicinity of an asbestos factory, the proportion quintuples, but even outside the heavily polluted areas in the city, the hazard is far higher than in the surrounding countryside.

TABLE No. 18

|  | Number of deaths | Number of inhabitants on 1.1.1965 | Number of mesotheliomas 10 000 inhab. | Per cent |
|---|---|---|---|---|
| Town of Bergedorf | 51 | 53 155 | 9.60 | 0.1 |
| Radius of 500 m around asbestos factory, i.e. area of approximately 0.785 km² (in Bergedorf) | 9 | 17 700 | 53.00 | 0.5 |
| Bergedorf without the 0.785 km² | 42 | 51 455 | 8.20 | 0.08 |
| Vierlande (agricultural suburb of Bergedorf) | 0 | 27 227 | 0 | 0 |
| Bergedorf district (Bergedorf + Vierlande) | 51 | 80 382 | 6.35 | 0.06 |

(c) Galloping pollution.

If it is true that asbestos creates a neoplasia hazard, the rapid growth in the quantity of asbestos extracted and utilised throughout the world should lead to an annual rise in the number of mesotheliomas diagnosed.

TABLE No. 19

|  | SOURCE | 1960 | 1961 | 1962 | 1963 | 1964 | 1965 | 1966 | 1967 | First half of 1968 |
|---|---|---|---|---|---|---|---|---|---|---|
| [34] | Ontario | 0 | 1 | 4 | 9 | 6 | 7 | 8 | 6 | 5 |
|  | Quebec | 6 | 6 | 4 | 3 | 8 | 7 | 13 | 14 | 8 |
|  | Other province | 10 | 2 | 5 | 7 | 3 | 5 | 10 | 6 | 2 |
| [37] |  | 8 | 2 | 4 | 4 | 11 | 11 | 9 | 20 |  |
|  | TOTAL | 24 | 11 | 17 | 23 | 28 | 30 | 40 | 46 | 15 |

## 3.2. Is asbestos responsible?

### 3.2.1. Hazards due to different types of asbestos

(1) **Chrysotile**

This does not appear to be dangerous [11]. Few mesotheliomas are reported from Canada, where it is extracted. It would entail a very long latent period. Nevertheless, several cases have been reported where the only exposure was to chrysotile.

(2) **Amphiboles**

(a) Antophyllite or amosite: few mesotheliomas when alone [20-29], and a longer interval also for:

(b) Crocidolite: the most suspect form.

- extraction:

The main source is Cape Province, South Africa. This is also the region where most mesotheliomas are encountered.

On the other hand, in the neighbouring province of Transvaal, where the crocidolite is extracted in conjunction with amosite, there are few mesotheliomas.

Yet working conditions are the same, and standards of medical supervision identical [33].

- utilisation:

Crocidolite was only imported into the USA from 1940 onwards [18]. Previously, Canadian chrysotile had been used.

Older workers have therefore been exposed to both forms.

However, in one plant where a careful study was made, no mesothelioma was diagnosed in any of the workers recruited after December 1942. In Australia, on the other hand, crocidolite has been held responsible for about ten cases [22].

It would appear that crocidolite is not solely to blame, but that it is a very suitable vector for one or more carcinogenic agents. The time factor, which will increasingly modify these data, should not be overlooked either.

### 3.2.2. Lesions recognised as being caused by or related to asbestos

The pathogenic affects of asbestos are well known:

1. Asbestosis is a pulmonary fibrosis, reactive to the presence of a foreign, irritant, indestructible body in the lung.

The pleural asbestosis-mesothelioma association is often mentioned [14, 26, 29, 30], these lesions being found in some 15 per cent of mesotheliomas.

2. The pleural plaques characteristic of asbestosis are also associated with a malignant evolutive pleural process. This pleural as well as pulmonary involvement is interesting.

3. Sero-fibrinous pleurisies remained without any etiology for a long time before it was considered that they were due to exposure to asbestos. These pleurisies have sometimes remained isolated, without evolving towards mesothelioma, and have even regressed without leaving a trace. Here, too, there has been pleural involvement.

### 3.2.3. The problem of asbestosic bodies

Asbestos fibre, a long thin particle, is breathed into the lung and remains there. The organism appears to react against this foreign body by covering it with a protein coating. However:

- this coating does not always exist
- it does not necessarily cover an asbestos fibre.

The first "asbestosic bodies" to be discovered by optical microscope sometimes turned out to be only pseudo-asbestosic bodies, with a core other than asbestos fibre. When colouring techniques revealed the presence of ferrous oxide in the coating (Perls positive ferro-protidic sheath), these bodies were usually called "ferruginous bodies".

But technical progress came to the help of pathologists, since the optical microscope has far too limited a power of separation to be able to identify the central fibre, which is several micrometres long but less than 1 µm in diameter.

The technique employed is X-ray diffraction, which makes it possible to determine the nature of the substance, followed by electronic micro-diffraction, which confirms the results of the first test by locating the particle with great precision [4]. Further improvements can be secured by using related techniques [2]; the electronic scanning microscope makes it possible to achieve a power of separation of 20 000 Å and to obtain a three-dimensional picture i.e. considerable depth of field. The electronic probe micro-analyser makes it possible to ascertain the nature of the crystal forming the backbone of the ferruginous body. The technicians hope to improve these methods still further.

The conclusions are extremely interesting, because it would appear that while the central fibre may of course be of asbestos (and the method makes it possible to determine the type of asbestos), it may equally well be of carbon, fibreglass or any other type of fibrous material.

The pollution problem needs to be reconsidered in the light of these new techniques.

(a) The presence of asbestos fibres in the atmosphere.

An investigation carried out in an asbestos factory has shown that many asbestosic particles which do not appear on the optical film of an air sample are discovered by electronic analysis of the same sample.

(b) Ferruginous bodies in man. Where are they to be found?

- In the sputum. It would appear that exposure must be very considerable for a positive result to be given. Stumphius in his thesis describes a systematic search for asbestosic bodies in the sputum of shipyard workers in Great Britain; he is rightly inclined to suspect that to some extent they may be ferruginous bodies.

TABLE No. 20

| OCCUPATION | Number of Sputa | Number of Asbestosic Bodies Ferruginous Bodies | Per cent |
|---|---|---|---|
| Shipyard | 87 | 65 | 75 |
| Engine shop | 88 | 52 | 59 |
| Other production shop | 40 | 22 | 55 |
| Other department (administration) | 32 | 8 | 25 |
| Asbestos workers (only ones subject to occupational exposure) | 24 | 22 | 92 |
| TOTAL | 271 | 169 | 62.5 |

- In the lungs: during a biopsy, exeresis or autopsy; by direct reading; by basal smear; by micro-incineration with activated oxygene, the asbestos being indestructible and therefore easily detected in the residue.

This method seems to be the most reliable [34].

- In the pleura: same methods.

The results vary widely according to the author and the method used. In cases of mesothelioma: asbestosic bodies, which often appear to be confused with ferruginous bodies, are discovered in the lungs of mesothelioma cases in varying proportions:

TABLE No. 21

| SOURCE | [1] | [13] | [15] | [21] | [25] | [26] | [27] | [30] |
|---|---|---|---|---|---|---|---|---|
| Number of mesotheliomas | 23 | 42 | 23 | 10 | 1 | 1 | 10 | 30 |
| Percentage of asbestosic bodies or ferruginous bodies in these mesotheliomas | 91 | 75 | 0 | 20 | 100 | 100 | 100 | 100 |

In routine autopsies:

TABLE No. 22

| SOURCE | [13] | [30] | [2] | [34] Profession | [34] Unexposed manual occupation | [34] Shipbuilding | [34] Women | [35] |
|---|---|---|---|---|---|---|---|---|
| Number of autopsies | ? | 200 | ? | 206 | 246 | 129 | 275 | ? |
| Percentage of asbestosic bodies or ferruginous bodies | 25 | 24 | 100 | 47 | 45 | 69 | 39 | 47 |

- In various cities of the world:

Meurman [50] reports the following figures, which are not strictly comparable, since they are arrived at by different authors and techniques.

TABLE No. 23

| YEAR | AUTHORS | COUNTRY OR CITY | PERCENTAGE |
|---|---|---|---|
| 1963 | Thompson | Capetown | 26.4 |
| 1965 | Elmes | Belfast | 14-27[1] |
| 1965 | Cauna | Pittsburg | 41 |
| 1966 | Hourihane | London | 24.3 |
| 1966 | Thomson | Miami | 27.2 |
| 1966 | Meurman | Finland | 57.6 |
| 1966 | Anjilvel | Montreal | 48 |
| 1967 | Roitzoch | Dresden | 43.2 |
| 1968 | Ashcroft | Newcastle | 53 |
| 1968 | Gibson | Glasgow | 51.5 |

[1] The percentage of Elmes varies from 14 to 27 between 1950 and 1969, depending on the year.

- In different occupations:

Goldstein [35] studies the question in the case of South African miners.

TABLE No. 24

| MINES | NUMBER OF CASES WITH FERRUGINOUS BODIES | PERCENTAGE |
|---|---|---|
| Gold | 89 | 19.7 |
| Coal | 25 | 5.5 |
| Asbestos | 250 | 55.3 |
| Other | 88 | 19.5 |

He also related the mesotheliomas to individual occupations and to the presence of ferruginous bodies:

TABLE No. 25

| MINES | MESOTHELIOMAS | | WITHOUT TUMORS | |
|---|---|---|---|---|
| | No Ferruginous Bodies | Ferruginous Bodies | No Ferruginous Bodies | Ferruginous Bodies |
| Gold | 1 | 0 | 7 858 | 89 |
| Coal | 0 | 0 | 1 201 | 25 |
| Asbestos | 0 | 8 | 89 | 250 |
| Other | 0 | 3 | 771 | 88 |
| TOTAL | 1 | 11 | 9 919 | 452 |

- With the years:

Whereas Elmes in Belfast [table 23] found an increase from 14 to 27 per cent of ferruginous bodies in autopsied lungs between 1950 and 1968, Selikoff [34] found no marked variations between lungs kept since 1934 and present-day lungs.

Yet the quantity of asbestos consumed today is far greater than in 1934. On the other hand, at that time, it was used virtually in the pure state, without taking precautions, whereas nowadays it is mixed with other elements. Far less dust is given off, but in view of the enormous increase in utilisation, the result seems to be the same.

## DISCUSSION

It would seem that where the figures differ, it is because the authors have employed different and usually inadequate or unreliable techniques, since electronic techniques appear to show that ferruginous bodies are ubiquitous in the lungs of modern man.

Moreover, asbestosic and ferruginous bodies have often been confused. Witwell [30] concluded from a more thorough investigation that the two might have been associated in 25 per cent of the 200 cases studied; in 8 cases out of 100, the ferruginous bodies were alone.

This ferro-protein coating appears to be due to a reaction by the body against the presence of an indestructible, very irritating and very sharp mineral fibre.

The hazard does not appear to come from these ferruginous bodies in other words, but from these bare fibres present in the lung and pleura, and even in the pleural plaques in certain cases of asbestosis.

New micro-electronic techniques have made it possible to determine with precision the composition of the coating and the nature of the central fibre.

This coating [35] appears to be composed of iron, phosphorous and calcium, in proportions similar to those of hemosiderin.

As regards the central fibre, it is possible to determine whether it is an asbestos fibre (in which case it is a genuine asbestosic body) and of what type, or alternatively, whether it is a fibre of other origin (fibreglass, carbon, etc.), in which case, it is a ferruginous body.

Pooley [36] accordingly concluded that:

(1) The lungs of sufferers from mesothelioma contain statistically more asbestos fibres than those of other people.

(2) Amphibole fibres are most often detected as bare fibres.

(3) The asbestosic bodies encountered are associated much more frequently with the presence of amphibole than chrysotile fibres.

(4) The asbestosic bodies are invariably formed around an amphibole core i.e. the asbestosic body is a reaction by the organism to the real danger represented by the presence of amphiboles.

These bare fibres remained undetected for a long time because of their infra-optical diameter.

But the electronic microscope has revealed them, and their presence forms the basis for widely varying theories on the neoplasia mechanism of the pleural cell.

### 3.2.4. The neoplasia of the mesothelial cell

Exposure of various animals to asbestos dusts has not led to any known cases of mesotheliomas, nor has the intra-tracheal injection of an asbestos solution [38, 39, 37].

On the other hand, the intra-pleural injection of this solution has given interesting results:

Wagner [40] has experimented with two types of rats: special rats raised in a sterile atmosphere free from any pathogenic hazard, and ordinary laboratory rats.

He injected them with different types of asbestos, especially crocidolite, first of all in the normal form and then in a form from which all traces of hydrocarbons had been extracted.

Lastly, a control group was given injections of a saline solution. There were no mesotheliomas in this group.

It was found that asbestos in any form causes a mesothelioma [38, 40, 41, 42, 43], chrysotile and crocidolite being the most dangerous.

Pure crocidolite does not appear to differ in its carcinogenic action from that extracted from oil (40), the risk of mesothelioma being roughly proportionate to the dose injected (table 26).

Stanton [44], after experimenting with rats to which he gave intra-pleural injections of different structural types of asbestos, glass fibre and aluminium oxide, three materials composed of fibres with a diameter ranging from 0.5 to 5 µm concluded that these fibres are more carcinogenic than bigger fibres or non-fibrous fibres of identical chemical composition.

The cancerogenous effect of these three fibres, and more particularly of asbestos in the case concerning us, would thus appear to be more directly related to their structure than to their physico-chemical properties.

TABLE No. 26

| Injection of a solution | | Amosite | Chrysotile | Crocidolite normal | without hydro-carbon | Salt |
|---|---|---|---|---|---|---|
| Special rats | TOTAL | 93 | 94 | 94 | 95 | 94 |
| | Mesothelioma | 38 | 61 | 55 | 56 | 0 |
| | Percentage | 41.3 | 64.9 | 58.5 | 59.5 | 0 |
| Normal rats | TOTAL | 83 | 89 | 88 | 87 | 81 |
| | Mesothelioma | 26 | 62 | 62 | 57 | 0 |
| | Percentage | 31.3 | 69.6 | 70.4 | 65.5 | 0 |

Asbestos has therefore been regarded as one carcinogenic factor among others, and has been studied from this standpoint by Schepers [45] in comparison or association with other substances.

Experiments with animals [46] do not reveal any difference between untreated asbestos and asbestos from which these hydrocarbons have been extracted; it is possible that insufficient time has elapsed for pathological manifestations to appear.

## REFERENCES

[1] Ashcroft, T., Heppleston, A.G., [1970]. Mesothelioma and asbestos on Tyneside: A pathological and social study. In: Pneumoconiosis, International Conference (Johannesburg, 1969), 177-179.

[2] Bignon, J., Goni, J., Remond, G. [1970]. Microscopie électronique à balayage et micro-analyseur à sonde électronique: application à l'étude des "corps ferrugineux" du poumon humain. Presse médicale, 78, no 19, 880-884.

[3] Bohlig, H., Hain, E. [1973]. Cancer in relation to environmental exposure. In: Biological Effects of Asbestos (Lyon, 1972), 217-221.

[4] Le Bouffant, Martin, Durif [1969]. L'identification des particules minérales dans les coupes histologiques par diffraction des rayons X. Le poumon et le coeur, 25, 3, 299-304.

[5] Champeix, J., Molina, C., Catilina, P., Cheminat, D.C. [1971]. Encyclopédie médico-chirurgicale, volume Poumons, Fascicule Pneumoconiose.

[6] Chretien, J., Delobel, J., Brouet, G. [1968]. Données étiologiques concernant 15 observations de mésothéliomes malins de la plèvre. Le poumon et le coeur, 24, 5, 549-557.

[7] Chretien, J., Delobel, J. [1968]. Mésothéliomes malins compliquant des séquelles de pneumothorax thérapeutiques. Journal français de médecine et chirurgie thoraciques, 22, 4, 383-403.

[8] Fourth International Pneumoconiosis Conference, Bucharest, 1971 [1973]. Apimondia Publishing House.

[9] Dalquen, P. [1970]. The epidemiology of pleural mesothelioma. A preliminary report of 119 cases from the Hamburg Area. German Medical, 15, 89-95.

[10] Delage, J., Mercier, R., Molina, Cl. [1968]. Les tumeurs pleurales primitives. A propos de 5 observations. Le poumon et le coeur, 24, 5, 505-518.

[11] McDonald, J.C., McDonald, A.D., Gibbs, G.W., Siematycki, J., Rossiter, C.E. [1971]. Mortality in the chrysotile asbestos mines and mills of Quebec. Archives of Environmental Health, 22, 6, 677-686.

[12] McDonald, A.D., Harper, A., El-Attar, O.A. and McDonald, J.C. [1970]. Epidemiology of primary malignant mesothelial tumours in Canada. In: Pneumoconiosis, International Conference (Johannesburg, 1969), Cape Town, Oxford University Press, 197-200.

[13] Elmes, P.C., Wade, O.L. [1965]. Relationship between exposure to asbestos and pleural malignancy in Belfast. Annals of the New York Academy of Sciences, 132,1, 549-557.

[14] McEwen, J., Finlayson, A., Mair, A., Gibbson, A.A.M. [1970]. Mesothelioma in Scotland. British Medical Journal, 4, 575-578.

[15] Gernez-Rieux, Ch., Voisin, C., Macquet, V., Leduc, M., Wallaert, C., Scherpereel, P. [1967]. Les mésothéliomes pleuraux diffus, expérience clinique de 12 années. Lille Médical, 12-4, 453-459.

[16] Gilson, J.C. [1970]. Asbestos Health Hazards: Recent observations in the United Kingdom. In: Pneumoconiosis, International Conference (Johannesburg, 1969), Cape Town, Oxford University Press, 173-176.

[17] Gracey [1971]. Pulmonary complications of asbestos exposure. Chest. 59, 1, 77-81.

[18] Selikoff, I.J., Hammond, E.C. and Churg, J. [1970]. Mortality experiences of asbestos insulation workers 1943-1968. In: Pneumoconiosis, International Conference (Johannesburg, 1969), Cape Town, Oxford University Press, 180-186.

[19] Jacob, K., Anspach, M. [1965]. Pulmonary neoplasia among Dresden asbestos workers. Annals of the New York Academy of Sciences, 132, 1, 536-548.

[20] Kiviluoto, R. and Meurman, L. [1970]. Results of asbestos exposure in Finland. In Pneumoconiosis, International Conference (Johannesburg, 1969), Cape Town, Oxford University Press, 190-191.

[21] Case records of the Massachusetts General Hospital [1971]. The New England Journal of Medicine, 284, 778-786.

[22] Milne [1969]. Fifteen cases of pleural mesothelioma associated with occupational exposure to asbestos in Victoria. The Medical Journal of Australia, 4 Oct., 669-673.

[23] Newhouse, M.L., Thompson, H. [1965]. Epidemiology of mesothelial tumours in the London area. Annals of the New York Academy of Sciences, 132, 1, 579-588.

[24] Schlienger, M., Eschwege, F., Blache, R., Depvere, R. [1969]. Mésothéliomes pleuraux malins. Etude de 39 cas dont 25 autopsies. Bulletin du cancer, 56, 3, 265-308.

[25] Turiaf, J., Basset, F., Battesti, J.P., Calvet, J.M. [1965]. Le rôle de l'asbestose dans la provocation des tumeurs malignes diffuses de la plèvre: "mésothéliome pleural". La presse médicale, 74, 39, 2199-2204.

[26] Turiaf, J., Chabot, J., Basset, F. [1968]. Cancer bronchique et mésothéliome pleural asbestosique (deux nouvelles observations). Le poumon et le coeur, 24, 5, 559-582.

[27] Vigliani, E.C., Mottura, G., Maranzana, P. [1965]. Association of pulmonary tumours with asbestosis in Piedmont and Lombardy. Annals of the New York Academy of Sciences, 132, 1, 558-574.

[28] Wagner, J.C. [1965]. Epidemiology of diffuse mesothelial tumours: evidence of an association from studies in South Africa and the United Kimgdom. Annals of the New York Academy of Sciences, 132, 1, 575-578.

[29] Webster, I. [1970]. Asbestos exposure in South Africa. In: Pneumoconiosis, International Conference (Johannesburg, 1969), Cape Town, Oxford University Press, 209-212.

[30] Whitwell, F., Rawcliffe, R.M. [1971]. Diffuse malignant pleural mesothelioma and asbestos exposure. Thorax, 26, 6, 6-22.

[31] McDonald, A.D., Harper, A., El Attar, O.A., McDonald, J.C. [1970]. Epidemiology of primary malignant mesothelial tumors in Canada. Cancer, 4, 26, 914-919.

[32] Hammond, E.C., Selikoff, I.J. [1973]. Relation of cigarette smoking to risk of death of asbestos-associated disease among insulation workers in the United States. In: Biological Effects of Asbestos (Lyon, 1972), 312-317.

[33] Harington, J.S. [1971]. Asbestos and mesothelioma in man. Nature, 232, 5305, 54-56.

[34] Selikoff, I.J., Hammond, E.C. [1970]. Asbestos bodies in the New York city population in two periods of time. In: Pneumoconiosis, International Conference (Johannesburg, 1969), Cape Town, Oxford University Press, 99-105.

[35] Catilina, P., Champeix, J. [1971]. Corps asbestosiques pulmonaires. Apport de la microsonde électronique de Castaing à la connaissance de la gangue et à l'identification de la nature asbestosique de la fibre centrale. Compte rendu des séances de la Société de biologie, T. 165, no 9-10, p. 1899.

[36] Pooley, F.D. [1973]. Mesothelioma in relation to exposure. In: Biological Effects of Asbestos (Lyon, 1972), 222-225.

[37] Behrens, W. [1951]. Über experimentelle Asbestosis. Schweizerische Zeitschrift für Pathologie, 14, 275-297.

[38] Wagner, J.C. [1962]. Experimental production of mesothelial tumours of the pleura by implantation of dusts in laboratory animals. Nature, 196, 180-181.

[39] Vorwald, A.J., Durkan, T.M., Pratt, P.C. [1951]. Experimental studies of asbestosis. Archives of Industrial Hygiene and Occupational Medicine, 3, 1-43.

[40] Wagner, J.C., Berry, G. [1969]. Mesothelioma in rats following inoculation with asbestos. British Journal of Cancer, 23. 1, 567-581.

[41] Smith, W.E., Miller, L., Churg, J., Selikoff, I.J. [1964]. Pleural reaction and mesothelioma in hamster injected with asbestos. Proceedings of the American Association for Cancer Research, 5, 59.

[42] Smith, W.E., Miller, L., Churg, J., Selikoff, I.J. [1965]. Mesotheliomas in hamster following intrapleural injection of asbestos. Journal of the Mount Sinai Hospital, 32, 1-8.

[43] Scheidemandel, V. [1971]. Asbestos and mesothelioma. Annals of Internal Medicine, June 1971, 1014.

[44] Stanton, M.F. [1973]. Some etiological considerations of fibre carcinogenesis. In: Biological Effects of Asbestos (Lyon, 1972), 289-294.

[45] Schepers, G.W.H. [1965]. Discussion. Annals of the New York Academy of Sciences, 132, 1, 504-506.

[46] Harington, J.S. [1965]. Chemical studies of asbestos. Annals of the New York Academy of Sciences, 132, 1, 31-47.

[47] Meurman, L.O., Hormia, M., Isomäki, M. and Rüttner, J.R. [1970]. Asbestos bodies in the lungs of a series of Finnish lung cancer patients. In: Pneumoconiosis, International Conference (Johannesburg, 1969), Cape Town, Oxford University Press, 404-407.

[48] Miller, L., Smith, W.E., Berliner, S.W. [1965]. Test of effect of asbestos on benzo(a)pyrene carcinogenesis in the respiratory tract. Annals of the New York Academy of Sciences, 132, 1, 489-500.

## 4. Bronchial Cancer and Asbestos

The statistics confirm the incidence of exposure to asbestos dust on the frequency of broncho-pulmonary cancer, and the pronounced character of this incidence.

### 4.1. Medical Literature

The literature on the subject is already extensive. The first cases appear to have been observed by Gloyne in 1933 and by Lynch and Smith in 1935.

In 1947, Mereweather (United Kingdom) found 31 cases of bronchial cancer in 247 autopsies of asbestosis cases i.e. 13.2 per cent.

In 1948, Lynch and Cannon (USA) reported three cases of bronchial cancer out of 40 autopsies of asbestosis cases.

In 1952, Behrens (Germany) found 44 cases of bronchial cancer in 309 autopsies of asbestosis cases i.e. 14.2 per cent (an over-all figure covering cases in Germany, USA, Canada and the United Kingdom).

In 1955, Cartier (Canada) observed 6 cases in 40 dead asbestosic workers (asbestos miners), but did not conclude that this frequency was particularly high because he found 7 cases of bronchial cancer among non-asbestosic workers.

In 1955, Hueper (USA) found 112 bronchial cancers in 738 autopsies of asbestosis cases i.e. 15 per cent.

In 1958, Chauvet (Geneva) published a new case, together with an interesting general review of the subject.

Also in 1958, Braun and Truan (Canada) made a general survey of 6 000 asbestos workers. In 187 autopsies, they found 9 definite cases of bronchial cancer, and 3 doubtful cases. These figures, when related to the number of workers exposed instead of to the number of deaths, gave a rate little higher than the rate of bronchial cancer for the province of Quebec and for Canada as a whole and even lower than for many urban areas in the USA.

In 1960, a case was published in a thesis by J. Hurel; Mrs. M.O. SAGOT in her thesis made a statistical analysis of the published cases.

In 1962, Cordova (USA) reported 11 new cases, but of varying origin.

Taken as a whole, these studies point to the conclusion that bronchial cancer is encountered in about 15 per cent of autopsies of sufferers from asbestosis as against only 0.2 - 2.78 per cent of all autopsies. This remarkably high frequency contrasts with the rarity of bronchial cancer in cases of silicosis.

The problem is quite complex because:

- these statistics overlap to an extent which is not always clearly stated; some cases have been published several times by different authors.

- all cases of asbestosis, especially in a discrete form, are not reported; all individuals dying of asbestosis are not autopsied, unlike virtually all cases of asbestosis and bronchial cancer.

The problem was investigated thoroughly by Bohlig, Jacob and Muller (Germany) in 1960. They covered, as precisely as possible, 57 cases of bronchial cancer occurring during asbestosis, in both parts of Germany, between 1939 and 1959. They encountered a number of major difficulties:

- the number of asbestos workers rose during the period in question from 3 000 to 10 000 and turnover was high.

- the number of cases of asbestosis notified during the same period was about 1 000 but the actual number of asbestoses was far higher, because many cancers occurred in very discrete asbestoses in respect of which no notification had been made or benefit granted.

- mortality from bronchial cancer among the population at large during the same period rose by 200 per cent among men and by 50 per cent among women.

- lastly, the risk of asbestosis declined during the same period as a result of improved protective measures.

Despite these difficulties, they concluded, after an exhaustive study of the cases, that bronchial cancer during asbestosis, compared with the total number of asbestos workers, is not more frequent than bronchial cancer in the general population of the same age, but that the risk of cancer in individuals suffering from asbestosis is markedly higher, perhaps by 100 per cent, than among the general population of the same age. Thus, quite independently, they reach the same conclusions as Braun and Truan.

## 4.2. Special Features of Cancer among Asbestos Workers

4.2.1. Latent period among individuals exposed to dust. It seems to be established by many foreign researchers that the interval between the beginning of exposure and the onset of cancer is between 20 and 30 years, in some cases as many as 40. It would also appear that whereas the intensity of the fibrosis and the number of serious respiratory disorders have fallen considerably since the introduction of preventive measures in the industry, broncho-pulmonary cancer has not diminished as a result. Finally, it is recognised that the degree of exposure to dust is not the only factor involved, and that the lapse of time since the first contact also counts, for once asbestos has entered the lungs, even in small quantitites, its pathogenic effects continue.

The observation made in Dresden which is mentioned by Selikoff in his 1967 editorial is interesting.

Before 1952, in 18 deaths from asbestosis, there was no cases of pulmonary cancer because these 18 deaths were due to severe pulmonary asbestosis with Cor pulmonale.

Subsequently, between 1953 and 1957, 9 bronchial cancers were found in 47 deaths. Between 1958 and 1964, 25 deaths out of 85 were due to pulmonary cancer and only 11 to pulmonary insufficiency.

This coincided with an improvement in prevention. Death from pulmonary asbestosis came after exposure for 25.7 years, whereas the interval between the start of exposure and death from lung cancer is 30.7 years.

With improved working conditions in the industry and less exposure to asbestos, the workers in the Dresden asbestos industry survived long enough to reach the age at which cancer supervened.

Of the cases described by Dousset in his thesis, the first, a woman, had always worked in the asbestos industry; her cancer started 34 years after she entered the industry. The second was still at work, after 31 years, when his cancer was detected. These two had thus spent their whole working lives in the industry. The third case involved a woman worker who had been employed on asbestos weaving for only 13 years, from 1934 to 1947, and her cancer occurred 32 years after the first exposure to dust.

With a latent period of 20-30 years, if not more, it is easy to understand how many workers who had changed jobs for one reason or another, how many women who gave up work on marriage, how others who left the industry because they could not stand the dust, have suffered or will suffer from cancer a long time later. It is only in exceptionally favourable conditions therefore that an industrial doctor is informed about these cases. Individuals who are subjected to intensive exposure and are fated to die of respiratory insufficiency, usually die before the onset of the cancer.

In other words, the length of the latent period is a source of error by default, but it is not the only one.

4.2.2. Sex

The incidence of pulmonary cancer among occupationally-exposed women is greater than among the general population.

Bauer, in a study of the causes of death of 128 men suffering from asbestosis, found a bronchial cancer in 17.2 per cent of cases; for 107 women, the corresponding figure was 8.4 per cent. Chauvet, in 30 observations of cancer among individuals exposed to dust, found 21 tumours in men, and 9 in women. Hueper found a comparable proportion of 2.3 men for 1 woman. Mancuso conducted 106 autopsies of exposed workers, and found 4 bronchial tumours in women.

Dousset et al., in 5 observations of cancer associated with asbestosis in the Pneumo-Physiology Department of the Le Havre hospital, noted that 2 of the cases involved women.

In the non-exposed population, however, bronchial cancer is 10 times commoner among men than among women. Strauli notes that out of 1 218 cases of bronchial cancer, 1 064 were men and 154 women i.e. 87.4 per cent of men and 12.6 per cent of women, or a little over 1 woman for 10 men. B. de Laguillaumie et al. concluded that 9 men are affected by cancer for 1 woman. The same proportion is suggested by Ochsner.

In other words, like smoking too much, exposure to asbestos dust tends to reduce the difference between the sexes in liability to broncho-pulmonary cancer.

The explanation probably lies in the fact that women are frequently employed in this industry; this also helps to confirm the carcinogenic effect of asbestos.

### 4.2.3. Topography of bronchial cancer in cases of asbestosis

This cancer tends to be found in the lower lobe of the lung. Out of 21 observations quoted by Chauvet, this location is found in 17. It is also encountered in the case described by B. de Laguillaumie and in the two observations by Rousselin et al. Our three observations conform to this pattern too. Similarly, Isselbacker found this location in 16 out of 20 cancer cases. Cordova et al. emphasises that several locations are possible.

Thus, location in the lower lobe predominates heavily in asbestosis, whereas it is by no means the rule in primary broncho-pulmonary cancer, in which there are twice as many neoplasms of the upper lobe. Delarue and Pailla find this cancer in the upper lobe in one case in two when in the right lung and in two cases out of three when in the left lung. Delmer, in a study of 1 457 cancers, found them located in the upper lobe in 56 per cent of cases and in the lower lobe in 33 per cent. Boyd also finds that the upper lobe is affected twice as often as the lower lobe by primary cancer.

### 4.2.4. Histology

Bronchial cancers during asbestosis do not appear to have any special histological features compared with other bronchial cancers. They have a very high proportion of epidermoid cancers, although some authors report a higher proportion of forms reaching maturity.

There is no relationship between the degree of seriousness of the asbestosis and the likelihood of cancer, which is just as probable if the asbestosis is discrete.

### 4.2.5. Multi-modular tumour

An important point is the possible multicentric origin of cancer during asbestosis. This has been observed by Lynch and Smith in the USA and by Bohmke in Germany.

Bohlig's statistics show 12 cases out of 33 parenchymatous tumours; it also occurred in our observation no. 1.

## 4.2.6. Relationship with the degree of evolution of the asbestosic fibrosis

This fibrosis has been considered by many authors to be the main as well as the initial cause of cancerisation. Subsequent observations tend to minimise its role. Undoubtedly, "the fibrosis comes first, but it is sometimes so discrete as to be undetected by radiological means" (J. Turiaf). O'Donnel asserts from personal experience that there is no relationship between the seriousness of the asbestosis and the development of the neoplasm. He has observed cases of cancer with few asbestosic bodies and other cases of major fibrosis in which death was due to asystolia, without any associated cancer. Selikoff shares this view and his observations in Dresden are mentioned elsewhere. The conclusion that would appear to emerge from the foregoing is that absence of a radiological indication of fibrosis or of asbestosic bodies in the sputum should not, in any general investigation, automatically rule out cases suffering from cancer which have also been exposed to asbestos.

## REFERENCES

Behrens, W. [1952]. Über Klinik und Pathologie der Asbestosis. Zeitschrift für Unfallmedizin, 45, 129-140 and 179-189.

Böhlig, H., Jacob, G., Muller, H. [1960]. Die Asbestose der Lungen: Genese-Klinik-Röntgenologie. Georg Thieme Verlag, Stuttgart.

Braun, O., Truan, P. [1958]. An epidermological study of lung cancer in asbestos miners. Archives of Industrial Health, 17, 634.

Cartier, P. [1955]. Some clinical observations of asbestosis in mine and mill workers. Archives of Industrial Health, March 1955, 204-207.

Champeix, J., de la Guillaumie, B., Jacquemet, J., Mory, F., Geille, A. [1962]. Cancer pulmonaire et exposition aux poussières d'amiante. Archives des maladies professionnelles, T. 23, 4-5, 267-271.

Chauvet, M. [1958]. Asbestose et cancer bronchique. A propos d'un cas. Presse médicale, 66, 908.

Cordova, J.F., Tesluk, H., Knudtson, K.P. [1962]. Asbestosis and carcinoma of the lung. Cancer, 15, 1181-1188.

Dousset, G., Desbordes, J., Tayot, J., Duwoos, H., Ernoult, J.L., Manouvrier, R., Veret, J. [1968]. Physionomie particulière du cancer bronchique chez les asbestosiques [A propos de trois observations nouvelles]. Le poumon et le coeur, T. 24, 5, 583-605.

Desbordes, J., Tayot, J., Dousset, G. [1968]. Réflexions sur l'effet carcinogène de l'amiante. Le poumon et le coeur, T. 24, 5, 607-619.

Dousset, G. [1968]. Le cancer broncho-pulmonaire de l'asbestosique [A propos de trois observations nouvelles recueillies à l'hôpital du Havre dans le courant de 1967]. Thèse Paris, J. Boileau, éditeur, Le Havre.

Hueper, W.C. [1955]. Silicosis, asbestosis and cancer of the lung. American Journal of Clinical Pathology, 25, 197.

Lynch, K.M., Cannon, W.M. [1948]. Asbestosis, analysis of forty necropsied cases. Diseases of the chest, 14, 874-885.

Nordmann, M., Sorge, A. [1941]. Lungenkrebs durch Asbeststaub im Tierversuch. Zeitschrift für Krebsforchung.

Rousselin, L., Ernoult, J.L., Veret, J., Manouvrier, F., Tayot, J. [1966]. Le cancer bronchique au cours de l'asbestose. Journal français de médecine et chirurgie thoraciques, 20, 37-51.

Selikoff, I.J., Bader, R.A., Bader, M.E., Churg, J., Hammond, E.C. [1967]. Asbestosis and neoplasia. American Journal of Medicine, 42, 487.

## TECHNICAL PREVENTION OF ASBESTOS HAZARDS

by

A. Wilkie, H.M. Chemical Inspector of Factories,
Department of Employment,
United Kingdom

### 1. Assessment of the Exposure

#### 1.1. Measurement of Airborne Dust Concentrations

The Hygiene Standards for asbestos in the United Kingdom referred to in this paper are based principally on a system of counting "fibres" with an aspect ratio of greater than 3.1 and which are greater than 5 µm in length. Only fibres which are visible by optical microscopy under transmitted light at a magnification of 450-500 times are counted and fibres with diameters greater than 3 µm or with adhering particles greater than 5 µm are currently excluded.

The membrane filter method of collection is the normal sampling system and positive phase-contrast microscopy is used after clearing the filter in order to improve the visibility of the fibres. Some of the earlier results on which the Hygiene Standard is based were achieved by collection of the airborne particles on the glass slides by thermal precipitation. In this case phase contrast is not needed to facilitate counting and the deposited fibres were counted dry. Field work has shown that there is little difference in number count between the two methods in spite of poor collection efficiency of large fibres when using a thermal precipitator but the added convenience of sampling methods which use membrane filter collection systems has led to their general adoption. We currently use Millipore RAWG gridded filters (1.2 µm pore size) and clear with triacetin.

The method of counting is based on a method agreed in detail with the Asbestosis Research Council representing the manufacturing industry and published in their Technical Note 1.

All "fibres" visible are noted and random fields are examined until at least 200 fibres have been counted or in the event of a sparse deposit 100 fields have been examined. A knowledge of the total volume of air aspirated enables a concentration to be expressed as fibres/$cm^3$.

On some occasions, where medical/environmental studies are being carried out, the gross figure of fibres/$cm^3$ is further split into fractions, e.g. fibres of length 5-9.9 µm, 10-19.9 µm, 20-50 µm and length greater than 50 µm all expressed as fibres in that range per cubic centimetre of air.

Other methods of sampling using impingement principles have not in our experience enabled quantitative counts to be derived which correlate the collection methods based on membrane filters or thermal precipitation.

Gravimetric methods have been used which depend on total airborne dust sampling or alternatively by collection of a "respirable" fraction after elutriation or cyclone separation. These were followed in each case by X-ray diffraction analysis to determine the type and percentage of asbestos in the dust sample.

Such systems are attractive in principle from an analytical standpoint but have not given consistent correlations with the fibre counting method. The separation of a respirable sample is inherently difficult and we do not know of any instrument yet marketed which is able to perform such a function on chrysotile asbestos fibre.

## 1.2. Principles_of_Sampling

Although background and snap samples are sometimes taken in order to evaluate the efficiency of dust collecting plant or the general operation of a process, all sampling intended to check compliance with legal requirements, that is to say assessment of a worker's exposure, is carried out by sampling in the breathing zone or by collecting a personal sample.

Breathing zone samples (i.e. within a 30 cm from the nose and towards the front of the face) are normally short period samples which will be taken over a 10-minute period by holding a filter holder with membrane filter, vertically orientated, adjacent to a worker engaged in a process to which the Asbestos Regulations apply. Inspectors will normally choose the sampling periods to coincide with periods of maximum dust evolution. The flow of air through the filter will be continuous or, alternatively, will be at regular intervals throughout the period of test in the event of manual aspiration being used. The size range of the particles collected for subsequent evaluation by optical microscopy will not be deliberately restricted.

Portable personal samplers are normally recommended and are in general used by HM Inspectors of Factories when sampling over long as well as short periods, as these are thought less likely to lead to errors in determining the dust concentrations and to give a better measurement of personal exposures. An appropriate filter holder may be worn on the lapel but other modifications such a head or shoulder harness are also likely to be accepted.

Time weighted personal exposures normally carried out in this instance over a 4-hour working period may be carried out using either a low flow rate personal sampling pump running continuously throughout this period or alternatively by taking an adequate number of representative samples and averaging these concentrations so as to represent a mean exposure over the 4-hour period; those separate samples will normally be of at least 1/2 hour duration.

Aspiration of air through the sampling head is normally achieved by using small battery powered mechanical pumps.[1] Sampling rates are variable up to a maximum of 2 litres per minute and the

---

[1] Such sampling pumps are available in the United Kingdom from Rotheroe and Mitchell Ltd., Perivale, Middlesex, or C.F. Casella Co. Ltd., Britania Walk, London N.1.

instrument will operate for 6-8 hours without recharging the battery. Frequent checks of flow rate during operation are essential.

Having completed the sample, the normal system of operation is to fix the dust deposit by spraying gently with a cytological fixative from an aerosol dispenser. The filter holder is held with the filter surface facing upwards and the aerosol dispenser held some 2 ft away so that the spray falls in an arc onto the membrane surface for a few seconds. Care has to be taken not to direct the aerosol spray on to the filter as this may disturb the dust deposit. A cover is then put over the filter holder so that the whole can be transported back to the laboratory for mounting and subsequent clearing for microscopy.

Light scattering instruments such as the Royce particle counter have been used for special applications. It is possible to relate the indicated size analyses and dust particle counts to fibre size analysis and counts determined by optical microscopy by regular calibration. They have the advantage of speed and assess dust particles in their undisturbed airborne condition. The principal use however is confined to textile processing workrooms where asbestos constitutes the major dust constituent.

### 1.3. Asbestos Regulations

When the injurious consequences of exposure to asbestos dust became apparent, the Asbestos Industry Regulations 1931 were introduced, which required the adoption of certain precautions aimed at reducing the exposure of the workers to the dust. It was to be expected that the energetic application of the preventive measures required by the Regulations would ensure a significant reduction in the incidents of asbestosis and while very great improvement has been achieved in those processes to which these Regulations applied, the use of asbestos in a great number of processes to which they did not apply gave rise to increasing concern. These processes were principally the application and removal of asbestos insulation and lagging. The Asbestos Regulations 1969 were introduced in May 1970 with wider application than the 1931 Regulations and, in addition, with more positive standards of dust control. Technical Data Note 13 (issued by HM Factory Inspectorate) gives an interpretation of concentrations of asbestos dust or dust containing asbestos which are liable to cause danger to health. The new Regulations apply to every process involving asbestos or any article composed wholly or partly of asbestos except a process in connection with which asbestos dust cannot be given off. The processes may be those occurring in factories, electrical stations, certain institutions, warehouses other than warehouses of any dock or forming part of any wharf or quay, ships in the course of construction or repair and on building operations or works of engineering construction (this includes demolition processes). For the purpose of the Regulations, asbestos is defined as any of the minerals crocidolite, amosite, chrysotile, fibrous anthophyllite or any mixture containing any of these minerals. In the Regulations references to asbestos dust are to be taken as references to dust consisting of/or containing asbestos to such an extent as to be liable to cause danger to the health of employed persons, who incidentally need not necessarily be persons employed on the actual asbestos process. The actual concentrations of asbestos dust which may be regarded as liable to cause danger is based on current information and is thus subject to review as further knowledge becomes available. It is at present established at 2 fibres/cm$^3$ or 0.1 mg/m$^3$.

## 2. Methods of Protection

### 2.1. Substitution

Substitution of asbestos by less toxic materials is the best method of eliminating the hazard and most of the large industrial organisations in the United Kingdom now demand non-asbestos lagging materials. These are generally of the magnesia or calcium silicate type reinforced with glass or man-made mineral fibre. Loose material, blankets and rope made of these fibres are also freely available. Ceramic fibre products are used for higher temperature applications.

Inspectors can only encourage substitution where it seems possible.

### 2.2. Dust Suppression

Safe conditions may be achieved when the asbestos in a process has been effectively wetted so that no dust at all is evolved in the course of the operations. Spray nozzles have been devised that can be inserted in lagging in such a manner that the asbestos layer is thoroughly soaked before it is removed. Certain asbestos compositions for pipe insulation can be effectively wetted within their bag containers before they are applied to the pipes. In certain processes the paper bags containing the asbestos are immersed in water, or steam is admitted to the contents before the bags are discharged into the hoppers of the process plant.

A considerable measure of dust control is achieved in certain textile processes by moistening the asbestos which is being processed. Similar satisfactory conditions may be present during the application of a spray or paste in which asbestos is incorporated in a substance like bitumen. When, however, the medium for a spray or paste is water or any volatile liquid, great care must be taken to ensure that drips or splashes do not in drying create a dust problem on their own. When the paste or spray is applied on site it is advisable to cover the floor in the vicinity with plastic sheeting or similar material. In some processes the use of a wet method is sufficient in itself to ensure safety, but in others it may have to be augmented by a certain measure of exhaust ventilation or by the provision of dust respirators and overalls.

The above considerations relate to processes from which dust would be evolved unless positive steps (wetting) were taken to prevent it, but there are some asbestos operations which are intrinsically dust free. The punching out of rubberised asbestos gaskets is a process which does not produce dangerous dust because the bonding effect of the rubber prevents the asbestos particles produced by the operations from escaping into the air in the form of dust. Progress has also been made in the development of some dust-suppressed asbestos textile materials which can be used without producing harmful quantities of dust.

## 2.3. Partial Enclosure with Exhaust Draught

The economics of asbestos production and the processes that are commonly associated with it do not readily lend themselves to total enclosure, but a similar type of enclosure is suitable for a number of operations to which asbestos is subjected, namely partial enclosure with exhaust draught. This can be a very effective means of preventing the dust caused by an operation from entering the air of a workroom. The nature of the process determines the general design of such an enclosure and the materials used in its construction, e.g. steel, wood, glass or plastic; but the common feature of all such enclosures is that the openings in them are specially planned to facilitate the passage of the work and the manipulation that has to be carried on inside them. Subject to these considerations the openings should be as small as possible and suction should be applied to the interior of the enclosure so that air moves inwards through the openings at sufficient speed to prevent an outward movement of dust. This air speed is related to the conditions which the process creates inside the enclosure.

Examples of such enclosure are found at opening plant, some bag emptying plant and some machining processes of small components incorporating asbestos fibre.

## 2.4. Booths

For the purposes of the work it may be necessary that the whole of one side of a ventilated enclosure should be open. A booth of this type therefore has a relatively large open area through which dust can escape if the design is faulty. The structure should be large enough to contain the dust cloud and deep enough to ensure that the eddies in the back corners caused by the exhaust suction will not spill out at the open face of the booth. If the asbestos process within a booth conforming to this design does not itself give rise to high pressures or turbulent conditions, the escape of dust will be prevented by an exhaust that produces an inward air movement of relatively low speed over the whole of the open face of the booth. On the other hand, if turbulence is caused inside the booth by high temperatures or by mechanical movement associated with the process, then higher inward air speeds will be required to counter the outward movement of the dust cloud. What is important is that the air speed should be adequate to control the conditions created by the process.

The correct positioning of the operator in relation to the dust source, therefore, will depend upon the air velocity into the hood, the design of hood and the speed at which the dust is released.

Typical processes to which this form of dust control can be applied are:

> charging bags of asbestos fibre to opening and mixing plant; bagging off asbestos fibre and dry mixtures containing asbestos; machining, fettling and sanding operations carried out on asbestos cement and resinated asbestos materials; and textile processes involving close handwork.

## 2.5. Receptor and Captor Hoods

When the process cannot be enclosed in the manner described in the preceding paragraphs, hoods, forming part of an exhaust ventilation system, should be fitted as near as possible to the source of the dust. These exhaust systems may be designed to receive and remove the dust-laden air that the process is delivering to them, or they may be designed to deflect dust particles that are moving in a direction imposed on them by the process, and extract them before they pollute the general atmosphere. Exhaust systems associated with receptor hoods must be capable of removing all the air that the process is delivering to the hood, and the fan of a captor system must draw air into the hood at a speed that is capable of changing the direction of the dusty air from the process. The speed of the air which is drawn into the hood falls off rapidly as the distance from the hood increases; the air speed into a circular, square or rectangular hood (but not a slot) at a point in space only one hood-face diameter from the centre of the hood face itself if only one-tenth of the speed at the face. Thus, the source of the dust should not be more than one hood diameter away from the hood if the ordinary exhaust system is to be effective. For the high velocities that abrasive wheels may impose on the dust particles extraction systems based on "low-volume high-velocity" principles have proved effective and should be considered if the methods described in the preceding two sections are impracticable. It should be emphasised, however, that the design of extraction systems for any particular process requires professional expertise and it should therefore be referred to a ventilating engineer of sound professional standing.

These methods have been applied with varying success to processes such as the following:

traditional textile processes;

machining and sewing of asbestos products;

most of the operations already mentioned, e.g. machining and mechanical sawing where it was thought that the particular process did not lend itself to a degree of enclosure.

## 2.6. Protective Equipment

When the circumstances are such that it is not practicable to comply with the requirements of the Regulations concerning exhaust ventilation, protective equipment must be provided for the use of each employed person exposed to asbestos dust liable to escape from a process (Regulation 8). The protective equipment must consist of approved respiratory protective equipment and suitable protective clothing.

*Respiratory protective equipment*. Where it is not practicable to control concentrations of asbestos dust to adequate standards, respiratory protective equipment will be necessary. This may take the form of an approved dust respirator except in high concentrations when face-piece leakage and other factors may make respirators ineffective. In these cases air-line breathing apparatus or pressure-fed respirators should be worn. Information about respiratory protective equipment approved by the Chief Inspector may be obtained from HM Factory Inspectorate. Reference may also be made to British Standards Institution publication BS 2091: 1969, "Specification for Respirators for Protection against Harmful Dust and Gases". It is essential that all such equipment is cleaned.

Air-line breathing apparatus. The Chief Inspector has approved "any breathing apparatus which consists of a properly fitting helmet or face-piece with necessary connections by means of which a person using it in a poisonous, asphyxiating or irritant atmosphere breathes ordinary air". The air may be pressure fed through a hose to the helmet or face-piece from a compressed-air system fitted with suitable control valves and an oil filter. Alternatively air may be pumped through the hose by means of a hand pump, or the wearer may be able to draw the air himself through a comparatively short hose. If one of the last two arrangements is used, great care must be exercised to ensure that the air entering the hose is not contaminated.

Positive-pressure powered respirators. Approved dust respirators have dust filters attached to the face-piece through which the wearer draws air in breathing and thus inhales air from which the dust has been filtered according to the efficiency of the filter. It is difficult to increase the efficiency of such respirators, however, in view of the leakage between the face-piece and the face of the wearer. This difficulty has been overcome in the positive-pressure, powered, dust respirators in which the air is driven through a filter to the face-piece by means of a battery-powered air blower carried on a belt by the wearer. There is thus a positive pressure which makes breathing easier and which counters any tendency of dust-laden air to leak into the face-piece at the edges in contact with the face. This respirator has many advantages for asbestos workers engaged in process, cleaning, repair and maintenance; for occasional work with crocidolite it should be regarded as representing the minimum standard. The pre-filter should be changed at least once per shift and the main filter checked periodically for clogging.

Protective clothing. It has been shown that the normal movements of the wearer of a dusty overall can dislodge the dust and disperse it in the air that he himself is breathing. It is manifestly desirable therefore that the material of the overall of the asbestos worker should be one that does not readily hold dust. Asbestos fibres adhere readily to the normal type of overall material, but a material consisting of 60 per cent polyester fibre and 40 per cent cotton has a relatively low tendency to become dusty and materials made wholly from synthetic fibres are even better in this respect. The overalls should be close fitting at the neck, ankles and wrists. The equipment should include a cap of the same material and in certain instances it may be augmented by plastic gloves, rubber boots and a plastic apron.

Choice of protective equipment. The nature of the protective equipment that should be provided is determined largely by the conditions of the work. It is only by atmospheric monitoring that a true assessment of relative exposures can be reliably made. When this is done it becomes possible to provide the workers with suitable protective equipment for the type and concentration of dust produced in his particular process or in his vicinity. Choice of respiratory protective equipment is governed principally by the concentrations of asbestos dust to which the operative is exposed as already stated. Further information is given in British Standards Institution publication BS 4275: 1968 "Recommendations for the Selection, Use and Maintenance of Respiratory Protective Equipment". It is expected that where crocidolite-containing insulation has to be exposed, it will not be sufficient to wear an approved ori-nasal respirator. Positive-pressure fed respirators or air-line breathing apparatus together with full protective clothing and associated facilities will be necessary.

## 2.7. Cleaning

Dust lying on floors, ledges and plant surfaces can become a source of air contamination when it is disturbed by draughts or by workroom activity and thus re-enters the atmosphere from which it originally settled. This settled dust can be a serious and persistent cause of harmful concentrations if it is not removed by efficiently directed cleaning. It has long been known that simple sweeping can prove to be an ineffective way of removing a dangerous industrial dust. It tends to waft the finest and most dangerous particles into the air from which it is deposited on more inaccessible ledges. The problem is not confined to asbestos, and equipment is available which enables cleaning methods to be applied which do not themselves raise dust.

*Vacuum cleaning plant*. Two forms of vacuum cleaning plant have been designed to cope with industrial conditions. The simplest form consists of a transportable unit mounted on a bogey which is wheeled to the place where it is to be used, the electric motor being connected with the electrical supply as a domestic vacuum cleaner is plugged into a house circuit. The filtered air is returned to the workplace where the cleaner is being used, and it is essential, therefore, that only high efficiency filters are fitted to vacuum cleaners being used for asbestos dust and that they are well maintained.

An arrangement of this type is suitable in situations where asbestos is used only at irregular and infrequent intervals; it could be used in a factory in which plant, that must be treated with asbestos, has from time to time to be installed or repaired. On the other hand, when asbestos is regularly used in the processes, the vacuum cleaning plant should preferably consist of a central suction, filtration and settling unit from which pipes run to these parts of the building in which vacuum cleaning is necessary. Openings with closure caps are fitted at intervals along these pipes so that the hose of a vacuum cleaning implement may be connected with the suction by inserting its adapter into an appropriate opening.

It is clearly desirable to avoid surfaces on which dust may settle. The walls of buildings in which asbestos dust is liable to be present should have smooth inner surfaces and rooms should be constructed in such a manner that there are as few ledges as possible on which dust may lodge.

## 2.8. Dust Collection and Filtration Plant

Asbestos dust removed from process by an exhaust air draught is in general conveyed in air suspension to a dry filter unit which may be preceded by a cyclone, or alternatively the dust is separated by a wet collector.

When the dust is collected dry, it should shake down into a receptacle for disposal such as a polyethylene bag which can readily be sealed.

Although filtration is required there are no standards laid down in the United Kingdom if the exhaust air goes to the atmosphere. If, however, the exhaust is recycled and returned to the workroom, H.M. Factory Inspectorate ask for efficient filtration such that the concentration of asbestos in the return air is less than 1/10 of the Hygiene Standard, i.e. 0.2 fibres/cm$^3$ excluding crocidolite. In the latter case, this is considered to be so dangerous that approval cannot be given to any recirculation scheme.

The Asbestos Regulations 1969 require this plant to be inspected at least once in every 7 days and thoroughly examined at least once in every 14 months. Prescribed particulars of test and the examination have to be maintained and these involve static pressures at all exhaust points, an investigation by either Tyndall beam lighting or dust sampling for leakages to the workplaces, and if the air is recirculated then dust concentration measurements in the return air need to be taken at the time of the thorough examination (Certificate of Approval No. AP1 made under the Regulations gives details of the prescribed particulars).

## 3. Problem Areas Remaining

These are known to lie chiefly in the use of asbestos insulation boards in the construction industry and delagging operations including demolition of old industrial installations. Asbestos spraying is also still suspect.

The reasons for the continuing difficulties are as follows:

(1) the itinerant nature of the workforce;

(2) the widely varying methods of work;

(3) delagging old industrial premises involves arduous work dealing with large quantities of asbestos, a considerable proportion of which is likely to contain crocidolite. There is also, therefore, a problem of identification;

(4) in many cases it is not practicable to control dust effectively and therefore personal protection has to be provided with all the attendant difficulties of ensuring its correct use;

(5) the acceptability of asbestos spraying is totally dependent on site labour (largely unsupervised) making the correct addition of water to the pre-damping cycle.

## REFERENCES

For United Kingdom references dealing with the control of asbestos dust and implementation of the Asbestos Regulations 1969, see p. 101.

## SURVEY OF STATUTORY PROVISIONS RELATING TO THE PREVENTION OF HEALTH RISKS DUE TO ASBESTOS

### 1. Introduction

This survey of national legislation and practice respecting the protection of workers against the risks resulting from occupational exposure to asbestos dust relates to a problem which, in view of the extensive and rapidly expanding use of this substance, has already been receiving attention from doctors, labour inspectors, and legislative authorities for some time past.

In some countries, the existence of diseases such as asbestosis (pneumoconiosis due to asbestos), cancer of the bronchi, pleura and peritoneum, and asbestos-induced skin corns, has led to the taking of action to prevent them, on the one hand, and to compensate workers suffering from them, on the other hand.

As more and more becomes known about the health risks attributable to asbestos dust, national legislative authorities are paying increasing attention to the preventive and protective measures required. However, not enough is yet known about less dangerous substitutes, statutory precautions to be observed in work with asbestos are laid down in only a few countries, and statutory rules defining standards of cleanliness in premises, ventilation and personal protection are still by no means widespread.

In some countries, specific rules for medical supervision have been laid down; in others, the legislation is more general, and does not refer specifically to the health hazards of asbestos.

Finally, it appears that a good deal remains to be done in the way of concerted action by the competent authorities, employers and workers to ensure that the increasing number of persons employed in this rapidly growing industry are properly informed and trained.

In the following sections, the most significant national rules and regulations designed to prevent exposure to asbestos dust and to protect workers against the risks inherent therein are briefly reviewed.

### 2. General

2.1. Asbestos is a broad term applied to a number of substances falling into two chief varieties, chrysotile and the amphiboles. These substances are natural calcium and magnesium hydrated silicates; they have a fibrous structure and are incombustible. Chrysotile asbestos (white asbestos) is a hydrated magnesium silicate found in serpentine rock. It is widely distributed in nature and accounts for some 93 per cent of the world's asbestos production. Amphibole asbestos varieties include amosite, crocidolite, anthophyllite, tremolite and actinolite. The last two substances have few industrial applications but can be mixed with natural talc to produce the commercial variety.

Asbestos fibres, if inhaled, can cause specific diseases such as asbestosis, cancer of the bronchi or mesothelioma of the pleura or peritoneum, etc. Spicules of asbestos, if they penetrate the skin, can produce horny indurations leading to callosities which have to be surgically removed. Hence any operation during which asbestos dust is given off can represent a potential health risk for numerous workers.

Many industrial operations entail a risk of exposure; they relate to:

(a) the mining of the ore, which is extracted with the surrounding rock, crushed and dried; the fibres (sometimes hammered by hand) are then sorted and collected in sacks;

(b) the use of the fibres to make asbestos textiles, conglomerates and other raw materials;

(c) the extremely numerous applications of asbestos, in the pure state or mixed with other substances, as a component of many heat insulants used in shipbuilding and building construction, and of a wide variety of industrial products, such as brake linings, filters of all kinds, gaskets and so on.

2.2. Hence the workers affected are, initially, those engaged in extraction. Asbestos is chiefly mined in Canada (at Asbestos and Thetford, in Quebec), in the Soviet Union (at Asbest, near Sverdlovsk), in the Republic of South Africa and in Rhodesia. It is also mined, but on a smaller scale, in Australia, China, Cyprus, Italy, Japan, Swaziland, Turkey, the United States and elsewhere.

Secondly, there are the workers engaged in the manufacture of textiles, conglomerates and other substances containing a proportion of asbestos. Such workers are to be found in large numbers in most countries where industry is well developed.

Finally, there are the even more numerous workers using substances containing asbestos, who are often ignorant of the risks to which they are exposed. They include, for instance, asbestos insulation workers in the building, construction or shipbuilding industries, and all the workers in many other industries (e.g. brake lining factories), who handle millions of tons of asbestos in every shape and form every year.

2.3. Only a few countries have adopted special rules and regulations respecting the protection of workers against the health hazards of asbestos. Nevertheless, every labour inspectorate and industrial safety institution is concerned with the elimination of occupational risks due to asbestos, their action being based in some cases on specific regulations already adopted, and in others as far as possible on general occupational health and safety standards which can be applied, _mutatis mutandis_, to the certain risks due to asbestos.

## 3. Scope of the Problem

3.1. Few definite figures have been published showing the frequency and severity of diseases due to asbestos. On the other hand, the problems of asbestos dust have been fairly extensively studied by medical circles in many countries; doctors are fully aware of the serious nature of these diseases, and considerable progress has been made in their detection. The International Occupational Safety and Health Information Centre (CIS)[1] has published a bibliography on the occupational hazards of asbestos, analysing more than a hundred documents published between 1963 and 1968. A list of reference material dealing with the same subject also appears in a study of asbestos published in the United States in 1969.[2]

3.2. The magnitude of the problems in question is also thrown into relief by figures for the world production of asbestos, which is increasing yearly. The figures given in the following table relate to 1966, and are taken from the United States study referred to above.

3.3. For a number of years now, some countries have possessed legislation referring to certain asbestos-induced diseases; in the last ten years, however, Australia (Queensland), Belgium, Denmark, the Federal Republic of Germany, Morocco, the Netherlands, Sweden and the United Kingdom have promulgated regulations of a more specific nature; these will be referred to in sections 4 and 5 hereinafter.

------------

[1] International Occupational Safety and Health Information Centre, c/o ILO, Geneva, Switzerland.

[2] US Department of Health, Education and Welfare: Preliminary Air Pollution Survey of Asbestos, a Literature Review. October 1969, 94 pp. 238 bibl. refs.

## World Production of Asbestos, 1966-1972
### (in thousand metric tons)

| Country | 1966 | 1967 | 1968 | 1969 | 1970 | 1971 | 1972 |
|---|---|---|---|---|---|---|---|
| WORLD[1] | 3 780 | 3 680 | 3 840 | 4 070 | 4 280 | 4 400 | 4 490 |
| Australia | 13 | 1 | 1 | 1[7] | 1 | 1 | 3 |
| Botswana | - | - | - | - | - | - | - |
| Brazil | 273 | 338 | 345 | ... | ... | ... | ... |
| Bulgaria[6] | 2 | 2 | 2 | 3 | 3 | 3 | 1 |
| Canada | 1 351 | 1 318 | 1 448 | 1 462 | 1 508 | 1 483 | 1 535 |
| China [3] | 140 | 150 | 150 | 160 | 170 | 160 | ... |
| Congo | 300 | 300 | ... | ... | ... | ... | ... |
| Cyprus | 25 | 18 | 17 | 19 | 26 | 25 | 27 |
| Czechoslovakia | 9 | ... | 17 | 23 | 28 | 28 | 33 |
| Egypt | 2[2] | 2 | 3 | - | 4 | 0 | 0 |
| Finland | 12 | 12 | 12 | 14 | 14 | 10 | 6 |
| France | - | - | - | - | - | - | - |
| Guyana | 40 | 33 | 27 | 20 | ... | ... | ... |
| India | 8 | 8 | 9 | 10 | 11 | 14 | 12 |
| Italy | 82 | 101 | 104 | 113 | 119 | 119 | 133 |
| Japan | 19 | 25 | 22 | 21 | 21 | 22 | ... |
| Korea, Rep. of | - | 2 | 3 | 6 | 1 | 2 | 5 |
| Namibia | 176 | 90 | 86 | 90 | ... | ... | ... |
| South Africa | 251 | 244 | 236 | 258 | 287 | 319 | 321 |
| Southern Rhodesia | 160[3] | 97 | 86 | 80 | 80 | 80 | ... |
| Swaziland | 36 | 35 | 36 | 36 | 33 | 36 | 34 |
| Turkey | 4 | 4 | 4 | 6 | 3 | 3 | 5 |
| USSR [3] | 755 | 769 | 816 | 962 | 1 066 | 1 152 | ... |
| United States[4] | 114 | 112 | 109 | 114 | 114 | 119 | 119 |
| Yugoslavia[5] | 8 | 9 | 10 | 11 | 12. | 15 | 11 |

Note: the figures refer to non-fabricated asbestos fibres and asbestos powder.

[1] Excluding Bolivia, Ethiopia, Democratic People's Republic of Korea, Madagascar, Philippines, Portugal and Romania.
[2] Beginning 1966, including vermiculite.
[3] Source: US Burea of Mines (Southern Rhodesia, beginning 1966).
[4] Asbestos sold or asbestos used by producers.
[5] Excluding asbestos powder.
[6] Asbestos fibres only.
[7] Beginning 1969, twelve months ending 30 June of year stated.

## 4. National Technical Provisions relating to Asbestos

### 4.1. Maximum concentrations

4.1.1. In a number of countries, the authorities have laid down figures for the maximum allowable concentration of asbestos dust in the atmosphere at places of work, and have published information in this connection.

4.1.2. Internationally, the Sub-Committee on Asbestos of the Permanent Committee and International Association on Occupational Health in September 1970 drew up recommendations concerning the assessment of the risks associated with asbestos fibres (air sampling, places and duration of sampling, fibre counting methods), as well as the determination of the physical and chemical characteristics and free silica content of different kinds of fibres and of any biologically active metallic or organic pollutants possibly present in dust samples. As regards the nature of fibres and the traces of metals or other substances present in them, and sampling points and periods, these recommendations refer to United States practice and to Technical Notes 1 and 2, published in the United Kingdom by the Asbestosis Research Council (see hereinafter).

4.1.3. In Canada, the Environmental Health Services Branch of the Ontario Ministry of Health, in Data Sheet No. 18 (March 1969) laid down that the limit for fibres of all kinds should be five fibres per cubic centimetre; Canadian national legislation, however, refers in more general terms to the levels adopted by the American Conference of Government Industrial Hygienists.

4.1.4. In the United States, the Federal Standards dated 2 June 1971 [10] authorise eight hours' work in an atmosphere containing not more than five asbestos fibres of size exceeding 5 µm per $cm^3$ of air (average figure) (from 1 July 1976 this figure will drop to two fibres). Fifteen minutes' work per hour for five hours (in an eight-hour day) will be authorised at levels of between five and ten fibres per $cm^3$. The maximum concentration, beyond which no work may be performed, is 10 fibres per $cm^3$ of air. Checks must be carried out once every six months at least (the method of checking concentrations is laid down).

4.1.5. In the United Kingdom, the Occupational Hygiene Society has generally recommended the following classification of the degree of exposure as a function of dust concentration:

| Concentration of dust | Average concentration over three months (fibres per $cm^3$) |
|---|---|
| Negligible | 0-0.4 |
| Slight | 0.5-1.9 |
| Medium | 2-10 |
| Heavy | Over 10 |

Other data, provided by Her Majesty's Factory Inspectorate, Department of Employment and Productivity [21] and by the Engineering Equipment Users' Association [36] make reference to the Asbestos Regulations 1969, as follows:

| Average concentration of asbestos dust | | Sample taken over | Advice |
|---|---|---|---|
| Number of fibres per $cm^3$ | Weight of fibres in $mg/m^3$ | | |
| Less than 2 | Less than 0.1 | 10 minutes | The 1969 Asbestos Regulations do not apply |
| Between 2 and 12 | Between 0.1 and 0.6 | 4 hours | An improvement in working conditions depends on concentration and exposure time |
| More than 12 | More than 0.6 | 10 minutes | The Regulations must be applied *in toto* |

It must be observed that for blue asbestos (crocidolite) the concentration is reduced to 0.2 fibres per $cm^3$ (0.01 $mg/m^3$). It should also be noted that, amongst other items, three technical memoranda containing recommendations have been issued in the United Kingdom by the Asbestosis Research Council [33, 34, 35] to facilitate implementation of the Asbestos Regulations 1969.

## 4.2. Especially hazardous operations

4.2.1. Operations regarded as involving a particularly severe asbestos hazard are not invariably listed. In practice, several national labour inspectorates consider that the most serious risks of contracting asbestos-induced diseases are caused by atmospheric suspensions of fibres less than 3 m in diameter and from 10-200 µm long, and hence that work in which such conditions obtain must be subject to very special supervision. Whenever an official list is drawn up, provision is invariably made for it to be kept up to date, for new products containing asbestos are constantly being brought into use.

4.2.2. In Morocco, an order dated 4 February 1960 included in a list of hazardous industrial processes operations which ordinarily entail a risk of exposure to the inhalation of asbestos dust; it included, for example, the drilling, excavation and extraction, as well as the dry crushing, grinding, filtering and the handling of asbestos ore or asbestos-bearing rock; the carding, spinning and weaving of asbestos fibres; the application of asbestos with sprayguns; the use of asbestos as a heat insulator, and the dry handling of asbestos in the manufacture of asbestos cement, asbestos-rubber joints, brake linings, and asbestos paper and cardboard.

4.2.3. Australia (Queensland) [4] and the United Kingdom [22] have adopted the same list of dangerous operations, as follows:

"1. Breaking, crushing, disintegrating, opening and grinding of asbestos, and the mixing of asbestos (including the mixing of asbestos with any other material) or the sieving of asbestos, and all processes involving manipulation of asbestos incidental thereto;

2. All processes in the manufacture of yarn or cloth composed of asbestos or asbestos mixed with any other material, including preparatory and finishing processes;

3. The making of insulation slabs or sections, composed wholly or partly of asbestos, and processes incidental thereto;

4. The making and repairing of insulating mattresses, composed wholly or partly of asbestos, and processes incidental thereto;

5. Any other process in the manufacture of articles composed wholly or partly of asbestos in connection with which asbestos dust is given off."

4.2.4. In Denmark, a notification dated 14 January 1972[1] forbids the use of asbestos or materials containing it for insulation from heat, noise or humidity, whether by application, brush painting or spraying. The Chief Labour Inspector can, however, authorise exceptions to this rule for the assembling or dismantling of such materials if the workers would incur no risk thereby.

4.2.5. Various bodies in the United States have issued instructions respecting the preventive health measures to be taken, including for example the wetting of asbestos heat insulation before dismantling, the use of hoppers with dust removal by exhaust ventilation when asbestos fibres are bagged, or mixed dry together or with other substances, the total enclosure of dusty processes, and the use of exhaust ventilation on fixed and portable tools for sawing, drilling or moulding products containing asbestos.

4.2.6. In April 1970, the city of New York issued special provisions respecting asbestos spraying, an operation regarded as especially hazardous, and similar provisions have since been introduced by other cities in the United States.

4.2.7. In 1967, in the United Kingdom, the Asbestosis Research Council published a recommended code of practice for the handling, working and fixing of asbestos and asbestos-cement products in the building and construction industries [35].

4.2.8. This research body has likewise published a series of pamphlets, some of which offer practical guidance in the safe handling of substances containing asbestos (such as asbestos-based friction linings).

4.2.9. In Sweden, a set of instructions respecting protection against occupational risks arising in work with asbestos was issued in March 1964 [37]; in it, the National Workers' Protection Administration gives general rules for the wetting of asbestos, the enclosure of machines and of continuous processing equipment, exhaust ventilation and other technical preventive measures.

------------

[1] See the Annex hereinafter, item 6.

## 4.3. Asbestos substitutes

4.3.1. Only in a few countries has any effort been made to replace substances with an asbestos base by substances not containing asbestos, or producing only a minimal amount of asbestos dust.

4.3.2. In the United Kingdom, the Asbestosis Research Council has embarked on studies of replacement products. Suggestions are also made by the Engineering Equipment Users' Association. The Department of Employment and Productivity, in a handbook issued in 1969 and revised in 1971 [36] provides some information about the technical specifications of certain substitute products, such as glass wool and various kinds of mineral wool.

4.3.3. In the Netherlands, the Dutch Labour Inspectorate has issued recommendations in this respect [19].

## 4.4. Authorisation to use asbestos

4.4.1. In very many countries, general provisions applying to industry indicate that no undertaking may operate without a permit delivered by the competent authorities if processes jeopardising workers' health take place therein. Thus, although this may not be specifically laid down, industrial processes involving the use of asbestos, such as the weaving of asbestos to produce fireproofing materials, may in practice be subject to prior authorisation.

4.4.2. The rules in force in Australia (Queensland) and the United Kingdom lay down that certain safety requirements must be met by buildings in which processes likely to produce asbestos dust take place; these requirements relate in particular to design and maintenance. Thus, such buildings must have floors and walls which are "smooth and impervious", and have "as few surfaces as is practicable on which asbestos dust can settle".

## 4.5. Storage and transport

4.5.1. In the United Kingdom and in Australia (Queensland), rules have been laid down governing the storage, warehousing and transport of loose asbestos. This must be kept in closed receptacles, and may be transported only in such receptacles, which must in addition be clearly marked. In the United Kingdom, details are given in recommendations made by the Engineering Equipment Users' Association [36].

4.5.2. In various countries, the transport of asbestos is subject to the general provisions respecting the transport of substances liable to give off hazardous dust.

## 4.6. Cleaning of premises

4.6.1. The more noxious the dust produced, the more important it is that premises should be regularly cleaned; in many countries this requirement is provided for under general regulations.

4.6.2. In Australia (Queensland) and the United Kingdom, the rules in force specify that all places or premises (e.g. buildings, ships, etc.) in which asbestos is used or worked shall be kept free from asbestos dust, by the use of vacuum-cleaning equipment or some other suitable method such that asbestos dust neither escapes nor is discharged into the air of any workplace. In the United Kingdom, the Asbestosis Research Council has issued recommendations on these matters [23], describing methods of cleaning industrial premises and of removing dust from machines and equipment as well as various types of fixed and portable vacuum cleaners, and even giving names of firms manufacturing equipment which is effective in removing asbestos dust.

4.6.3. In the United States, the rules in force also provide for premises to be kept clean, and describe how waste should be removed. Any accumulation of asbestos fibres which if scattered might produce excessive atmospheric concentrations of asbestos dust must be removed from any surface on which it may have formed at workplaces. In addition, more generally, all asbestos wastes, contaminated clothes and objects which might emit dangerous quantities of dust must be stowed in sealed containers.

4.7. Ventilation

4.7.1. Many countries have issued general regulations calling for the general ventilation of working premises by exhaust ventilation and fresh air intake, together with local ventilation of individual workplaces. These measures are of especial importance for the elimination of asbestos dust in suspension in the atmosphere.

4.7.2. Further details in this respect are sometimes given in specific texts. This is, for instance, the case in Canada (for all mines), and in most of the Canadian provinces (for certain workplaces). In addition, provision is made in some Canadian provinces for water-spraying as a means of asbestos dust abatement, especially in mines.

4.7.3. In the United States, vigorous encouragement is given to the use of exhaust ventilation equipment for dust removal, for it is felt that dust masks affording adequate protection are hampering and likely to be discarded. Apart from equipment for keeping the workplace air clean, appropriate devices for asbestos dust removal must be fitted on machine tools used to work substances containing asbestos (saws, grinding and drilling machines, etc.) and on tables and workbenches where substances likely to emit such dust are unpacked. Dust abatement by wetting is urged whenever this is possible. The dust concentrations in the air of workplaces must not exceed the limits set.

4.7.4. In the United Kingdom, all operations during which asbestos dust is liable to be emitted are subject to the Asbestos Regulations 1969, which lay down that dust concentrations shall be kept below prescribed limits, largely by ventilation. The same holds good in Australia (Queensland) [4]. These two sets of regulations in addition lay down that ventilation equipment shall be inspected once a week and given a thorough overhaul every fourteen months. In the United Kingdom, the Department of Employment and Productivity, the Asbestosis Research Council and the Engineering Equipment Users' Association all provide further information.

4.7.5. Sweden has also issued instructions providing for the wetting of asbestos and the removal of asbestos dust by exhaust ventilation equipment.

4.8. Personal protective clothing and equipment

4.8.1. In very many countries[1], provision is made for the supply of personal protective clothing and equipment to workers. In some of them, it is specified that such clothing and equipment must be provided, cleaned and maintained at the employer's expense.[2] Often enough, such provisions are backed by general measures designed to ensure that the worker is not exposed to harmful substances. Lastly, some countries (instances are quoted below) have promulgated provisions relating specifically to the occupational hazards of asbestos.

4.8.2. In the United States, personal protective equipment must be worn in all demolition work and in asbestos spraying. Such equipment must not be used for more than a certain time. More generally, depending on the dust concentration levels observed at workplaces, the use of either plain dust masks or of supplied-air respirators with hoods is recommended. Fuller equipment must in certain circumstances be worn. Workers whose clothes are contaminated with asbestos dust must be provided with special changing rooms, and protective equipment must be regularly cleaned by competent persons.

4.8.3. Detailed instructions are given in the United Kingdom Asbestos Regulations 1969 and in the Australia (Queensland) Asbestos Rule of 9 July 1970. These provide, amongst other things, that personal protective equipment must be kept in good condition and be thoroughly cleaned and disinfected before being made over to another worker; moreover, workers must be properly instructed in its use. Special cloakrooms for changing and storage of clothes must be provided. In the United Kingdom, HM Factory Inspectorate has issued a Technical Data Note No. 24 [22] on respiratory protective equipment for use with various concentrations of chrysotile asbestos, amosite, fibrous anthophyllite and blue asbestos. Further detailed instructions on the use, upkeep, cleaning, etc., of personal protective equipment have been issued by the Department of Employment and Productivity, the Asbestosis Research Council and the Engineering Equipment Users' Assocation.

4.8.4. In Sweden, the 1964 Instructions give detailed rules for the use of respiratory protective devices.

---

[1] Among others: Austria, Belgium, Brazil, Cameroon, China, Costa Rica, Cyprus, Finland, France, Germany (Federal Republic of), Hungary, India, Iraq, Italy, Japan, Kenya, Malawi, Mali, Mexico, New Zealand, Romania, Spain, Singapore, Sri Lanka, Sweden, Switzerland, the Ukrainian Soviet Socialist Republic, the Union of Soviet Socialist Republics, the United Kingdom, the United States and Yugoslavia.

[2] Among others: Austria, Belgium, Bulgaria, Denmark, Finland, Hungary, Iran, Italy, Japan, Kenya, Mali, Mexico, Morocco, Norway, Poland, Romania, Spain and Switzerland.

## 4.9. Employers' and workers' responsibilities

4.9.1. Many countries[1] have adopted provisions sanctioning the legal principle that employers are responsible for protecting the health of their workers at their places of work. In numerous instances this responsibility is not limited to the technical aspects of the work and the provision of protective equipment, but also extends to safety and health training, medical supervision and so on.

4.9.2. In Australia (Queensland) and the United Kingdom, the employer must ensure that the law is applied in all operations during which asbestos dust may be emitted, unless he has first obtained special written authorisation from the chief inspector of factories.

4.9.3. There are many countries, too, in which the regulations lay down that the worker must make proper use of the protective devices provided for him. The Australia (Queensland) Asbestos Rule already cited makes special reference to the worker's responsibilities in this connection.

4.9.4. In the United States, the employer is responsible for installing, operating and maintaining equipment designed to prevent asbestos dust concentrations in the atmosphere from becoming dangerous, and in case of need it is he who has to supply the staff with the requisite personal protective equipment. He remains responsible even in respect of work given to outsiders. Thus, for instance, if he gives clothes contaminated with asbestos to outsiders to clean, he is under an obligation to draw their attention to the risks involved.

4.9.5. In Sweden, the regulations lay down that the employer is responsible for the protection of the health of workers exposed to asbestos dust.

## 5. Medical supervision

### 5.1. Medical examinations

5.1.1. In a number of countries, the national legislation lays down that any worker employed in the extraction of asbestos, the manufacture of products containing asbestos, or the use, application or working of such products, must, if the processes involved are liable to produce asbestos dust, be given a medical examination before such employment and thereafter at appropriate intervals, supplemented by X-ray and biological examinations, if required, so as to keep a check on his health and on the degree of exposure.

5.1.2. In Belgium, the Association of Belgian Industrialists has issued a publication [5] on the pre-employment and periodical

------------

[1] They include Belgium, Bolivia, Bulgaria, Cameroon, Chile, Colombia, Costa Rica, Czechoslovakia, Egypt, France, Guinea, Panama, Romania, Tunisia, Turkey, the Soviet Union, Venezuela, Viet-Nam and Zaire (International Labour Conference, 1970, Report III, Part 4, pp. 252-253).

medical examinations to be performed on workers in contact with asbestos dust. It refers to the general national legislation in force in that country.

5.1.3. In Canada, pre-employment and periodical examinations are provided for, notably for underground mineworkers (surface workers are also covered in some provinces). Workers in all industries using asbestos must be so examined (as in Alberta, for example), including workers engaged in applying and removing heat-insulating materials. In British Columbia, a worker must be medically examined within one month from his recruitment, and thereafter once a year, in Manitoba, every sixty days after recruitment and thereafter annually, and in Alberta, every two years; the examination must be supplemented by a lung X-ray.

5.1.4. In the United States, any worker occupationally exposed to asbestos dust must, within thirty days of being assigned to such work, undergo a medical examination which must include at least a lung X-ray, medical history and lung function tests (forced vital capacity and forced expiratory volume). He must also undergo an identical examination once a year and within thirty days after leaving his job. No further examination need to be given if the results of an examination undergone during the previous year are to hand. The results of all the examinations must be kept on file for twenty years.

5.1.5. In Morocco, a decree dated 2 February 1960 [14] lays down the medical precautions to be taken in undertakings where the staff are normally exposed to free silica or asbestos dust, and defines the medical examinations to be undergone on recruitment and at intervals thereafter. All examinations must include either a lung X-ray or radiograph obtained by a process approved by the Ministry of Public Health. On recruitment, a general clinical examination is given, and this is also required should an X-ray detect some anomaly on the occasion of a periodical examination. The worker must be medically examined at least once a year. An order dated 4 February 1960 [16] gives a list (see 4.2.2.) of jobs calling for the above-mentioned medical precautions. An order dated 5 February 1960 [17] lays down standards to be observed in such examinations. Another order, dated 6 February 1960 [18], makes recommendations concerning the X-ray equipment to be used, clinical examinations, pre-employment and periodic examinations (occupational history, counter-indications - notably condition of lungs - medical history, whether alternative work is available, and so on).

5.1.6. In the Federal Republic of Germany, the Federal Ministry of Labour and Social Affairs in 1971 issued rules governing medical examinations to be undergone by workers exposed to harmful dust of mineral origin [2], and in 1971 also, the Federation of Industrial Mutual Accident Insurance Assocations issued a set of principles applicable to preventive medical examinations [3]. Under regulations issued in 1964 concerning workers exposed to a risk of asbestosis, such workers must (if the Labour Protection Commission so requires) undergo medical examination in accordance with the procedures set forth in the general instruction respecting medical examinations issued in 1949.

5.2. Protection of young workers

5.2.1. As a general rule, young workers are protected by regulations dealing with jobs deemed harmful to the health of adolescents. Such regulations exist in very many countries, although they do not always refer specifically to work involving exposure to asbestos dust.

5.2.2. Some countries, nevertheless, do possess special rules. Thus, in the United Kingdom, as well as in Australia (Queensland), no young person may be employed in any process in which asbestos dust is liable to escape, or in carrying out any cleaning, other than by means of suitable vacuum-cleaning equipment. In Sweden, young workers may not be employed on work involving exposure to asbestos if there is a risk of contracting asbestosis.[1]

5.3. Compensation of diseases caused by asbestos

5.3.1. From 1930 onwards, asbestosis has been recognised as the major cause of incapacity among workers in the asbestos textiles industry in the United Kingdom, and subsequently in many other countries. Thus, there are many countries whose legislation already includes asbestos-induced cancer in the schedule of occupational diseases.

5.3.2. In the United Kingdom the National Insurance (Industrial Injuries) Amendment Regulations, 1966, enlarge the coverage of the diseases due to asbestos specified in the Regulations of 1959, by adding to the asbestosis the diffuse mesothelioma of the pleura or of the peritoneum.

5.3.3. In the United states, the asbestos regulations provide for the keeping of records showing the results of medical examinations performed in accordance with the relevant rules. Such data have to be kept on file for twenty years. The Assistant Secretary of Labor for Occupational Health and Safety, the Director of the National Institute for Occupational Safety and Health, and, on request, the employer concerned and doctors approved by them can consult these files and take notes.

5.3.4. In Morocco, an order dated 3 February 1960 [15] lays down procedures for the reporting of cases of occupational silicosis and asbestosis (investigation, autopsy, medical opinions), the granting of daily allowances, the award of pensions, the determination and review of reassignment grants, and medical check-ups of workers leaving jobs involving exposure to asbestos dust.

6. Information and training

6.1. In various countries, the competent authorities and the occupational safety and health organisations encourage research and disseminate information on asbestos hazards. This is the case in

---

[1] Royal Proclamation to Prohibit the Employment of Young Persons on Certain Dangerous Types of Work. No. 209 (Svensk Författningssamling, 19 May 1949, p. 413) (ILO: Legislative Series, 1949 - Swe. 3).

particular in Australia (Queensland), Belgium, the Federal Republic of Germany, the Netherlands, Sweden and, especially the United Kingdom and the United States, where numerous publications have been issued on this subject.

6.2. Employers are sometimes asked by insurance organisations to acquaint themselves with the occupational risks of asbestos dust, in connection with the manufacture or use of products containing asbestos; this occurs in particular in countries where insurance premiums vary with the kind of work carried on.

6.3. There are often rules obliging employers to keep their staff fully informed about the precautions to be taken against potential occupational hazards. Such special provisions exist, for example, in Sweden, where employers must see to it that foremen and workers working with asbestos are informed of the attendant risk of asbestosis and of the precautions to be taken. Swedish employers must give all necessary instructions in this respect. These instructions sometimes take the shape of permanent warning notices, such as, for example, posters carrying the message that a worker exposed to asbestos dust is more likely to contract bronchial cancer if he is a cigarette smoker than if he is not.

## Selected Special Instructions and Recommendations relating to Asbestos

### International

[1] Evaluation of Asbestos Exposures in the Working Environment - Recommendation by the Sub-Committee on Asbestosis of the Permanent Commission and International Association on Occupational Health. Work - Environment - Health (Helsinki, Finland, 1971), Volume 8, No. 3, pp. 71-73.

### Federal Republic of Germany

[2] Instructions concerning the Medical Examination of Workers exposed to the Effects of Harmful Mineral Dusts (communication by the Federal Ministry for Labour and Social Affairs), Bonn, 23 March 1971. Arbeitsschutz (Cologne), April 1971, No. 4, pp. 106-107.

[3] Principles applicable to Preventive Medical Examination - Risks associated with Harmful Mineral Dusts. Federation of Industrial Mutual Accident Insurance Associations, Bonn, January 1971. Arbeitsmedizin - Sozialmedizin - Arbeitshygiene (Stuttgart), April 1971, Volume 6, No. 4, pp. 94-95.

### Australia (Queensland)

[4] The Asbestos Rule, Department of Industrial Affairs, Brisbane, Queensland. Promulgated by Order in Council on 9 July 1970. Took effect on 11 July 1971. Queensland Government Gazette, Brisbane, Queensland, Volume 74, pp. 314-315.

### Belgium

[5] Poussières d'amiante. Note No. 15B, Association des industriels de Belgique, Brussels, Belgium, November 1960.

### Denmark

[6] Notification No. 18 (14 January 1972) forbidding the use of asbestos in certain kinds of insulation. Lovtidende A, 1972, No. II, Text No. 18, p. 24.

## United States

[7] Occupational Safety and Health Act of 1970, Public Law 91-956, 91st Congress, S. 2193. Approved on 29 December 1970. Took effect on 28 April 1971. (Also published in Spanish, English and French in ILO: Legislative Series, 1970 - USA 1, November-December 1970. 45 pp.)

[8] Implementation Regulations of Occupational Safety and Health Act of 1970. 29 Code of Federal Regulations, Chapter 17, Part 1910.

[9] Standard for Exposure to Asbestos Dust in Ship Repairing, Shipbreaking and Longshoring. Modification of paragraphs 1910.13 to 16 and 1910.93 a. Federal Register, Washington, DC, No. 6, 11 January 1972, p. 332.

[10] Occupational Safety and Health Standards - Standard for Exposure to Asbestos Dust. Code of Federal Regulations, Title 29, Chapter 17, Part 1910, Department of Labor, Washington, DC. Promulgated on 2 June 1972. Took effect on 7 July 1972 (certain sections on 1 July 1976). Federal Register, Washington, DC, 7 June 1972, Volume 37, No. 110, pp. 11318-11322.

[11] Occupational Safety and Health Standards - Asbestos. Code of Federal Regulations, Title 29, Chapter 17, Part 1910, paragraph 1910.93 a. Department of Labor, Washington, DC. Federal Register, Washington, DC, 18 October 1972, Volume 37, No. 202, p. 22144.

[12] Criteria for a Recommended Standard - Occupational Exposure to Asbestos. National Institute for Occupational Safety and Health, Public Health Service, Department of Health, Education and Welfare, Cincinatti, Ohio, United States, 1972. 136 pp.

## Italy

[13] Act No. 455 of 12 April 1943 to extend compulsory insurance against occupational diseases to cover silicosis and asbestosis, together with the medical supervision of staff at the employer's expenses. Gazzetta Ufficiale, 14 June 1943, No. 137, pp. 2066-2068.

## Morocco

[14] Decree No. 2-59-0219, dated 2 February 1960, determining special medical precautions to be taken in undertakings where workers are normally exposed to free silica or asbestos dust. Présidence du Conseil, Rabat. Took effect on 19 April 1960. Bulletin Officiel, Rabat, 19 February 1960, Volume 49, No. 2469, pp. 383-384.

[15] Order of the Minister of Labour and Social Affairs, dated 3 February 1960, setting forth rules for the application of legislation concerning the compensation of occupational diseases to silicosis and asbestosis of occupational origin. Bulletin Officiel, Rabat, 19 February 1960, Volume 49, No. 2469, pp. 387-390.

[16] Joint Order of the Minister of Labour and Social Affairs and the Minister of the National Economy, dated 4 February 1960, giving a list of industrial work normally exposing workers to the inhalation of free silica or asbestos dust. Bulletin Officiel, Rabat, 19 February 1960, Volume 49, No. 2469, p. 384.

[17] Joint Order of the Minister of Labour and Social Affairs, the Minister of National Economy and the Minister of Health, dated 5 February 1960, making recommendations to doctors responsible for supervising staff exposed to silicosis and asbestosis hazards. Bulletin Officiel, Rabat, 19 February 1960, Volume 49, No. 2469, pp. 385-386.

[18] Joint Order of the Minister of Labour and Social Affairs, the Minister of National Economy and the Minister of Health, dated 6 February 1960, making recommendations concerning the X-ray equipment to be used in detecting and following up cases of silicosis and asbestosis. Bulletin Officiel, Rabat, 19 February 1960, Volume 49, No. 2469, pp. 386-387.

## Netherlands

[19] Werken mit Asbest (Work on asbestos). P. No. 116, Arbeidsinspectie, Directoraat-Generaal van de Arbeid, Voorburg. First edition, 1971. 14 pp., illus.

## United Kingdom

[20] The Asbestos Regulations 1969, Department of Employment and Productivity, London. Promulgated on 13 May 1969. Took effect on 14 May 1969. Statutory Instruments, 1969, No. 690. Her Majesty's Stationery Office, PO Box 569, London SE1, England.

[21] Standards for Asbestos Dust Concentration for Use with the Asbestos Regulations 1969. Technical Data Note 13, HM Factory Inspectorate, Department of Employment and Productivity, London, England. Her Majesty's Stationery Office, PO Box 569, London SE1. No date. 3 pp. Free of charge.

[22] Asbestos Regulations 1969: Respiratory Protective Equipment. Technical Data Note 24, HM Factory Inspectorate, Department of Employment and Productivity, London. Her Majesty's Stationery Office, PO Box 569, London SE1. July 1971. 2 pp.

[23] Protective Equipment in the Asbestos Industry (Respiratory Equipment and Protective Clothing). Control and Safety Guide No. 1 (revised), Environmental Control Committee, Asbestosis Research Council, 114 Park Street, London W1Y 4AB. November 1971. 12 pp., illus.

[24] The Application of Sprayed Asbestos Coatings. Control and Safety Guide No. 2 Environmental Control Committee, Asbestosis Research Council, 114 Park Street, London W1Y 4AB. March 1970. 8 pp.

[25] The Stripping and Fitting of Asbestos-Containing Thermal Insulation Materials. Control and Safety Guide No. 3, Environmental Control Committee, Asbestosis Research council, 114 Park Street, London W1Y 4AB. March 1970. 8 pp.

[26] Asbestos-Based Materials for the Building and Shipbuilding Industries and Electrical and Engineering Insulation. Control and Safety Guide No. 5, Environmental Control Committee, Asbestosis Research Council, 114 Park Street, London W1Y 4AB. March 1970. 8 pp.

[27] The Handling, Storage, Transport and Discharging of Asbestos Fibre into Manufacturing Processes. Control and Safety Guide No. 6 (revised), Environmental Control Committee, Asbestosis Research Council, 114 park Street, London W1Y 4AB. December 1971. 11 pp., illus.

[28] The Control of Dust by Exhaust Ventilation when working with Asbestos. Control and Safety Guide No. 7, Environmental Control Committee, Asbestosis Research Council, 114 Park Street, London W1Y 4AB. April 1970. 11 pp., illus.

[29] Asbestos-Based Friction Materials and Asbestos-Reinforced Resinous Moulded Materials. Control and Safety Guide No. 8, Asbestosis Research Council, 114 Park Street, London W1Y 4AB. December 1970. 8 pp.

[30] The Cleaning of Premises and Plant in accordance with the Asbestos Regulations. Control and Safety Guide No. 9, Environmental Control Committee, Asbestosis Research Council, 114 Park Street, London W1Y 4AB. 1972. 11 pp.

[31] Recommended Code of Practice for the Handling and Disposal of Asbestos Waste Material. Asbestosis Research Council, 114 Park Street, London W1Y 4AB. September 1969. 8 pp.

[32] Recommended Code of Practice for handling Consignments of Asbestos Fibre in British Ports. Asbestosis Research Council, 114 Park Street, London W1Y 4AB. February 1972. 4 pp.

[33] The Measurement of Airborne Asbestos Dust by the Membrane Filter Method. Technical Note 1, Asbestosis Research Council, Dr. S.P. Holmes, Secretary, PO Box 40, Rochdale, Lancashire, England. New edition revised September 1971. 10 pp., illus.

[34] Dust Sampling Procedures for use with the Asbestos Regulations, 1969. Technical Note 2, Asbestosis

Research Council, Dr. S.P. Holmes, Secretary, PO Box 40, Rochdale, Lancashire. January 1971. 7 pp.

[35] A Recommended Code of Practice for Handling, Working and Fixing of Asbestos and Asbestos-Cement Products in the Building and Construction Industries. Asbestosis Research Council, PO Box 40, Rochdale, Lancashire. April 1967. 4 pp.

[36] Recommendations for Handling Asbestos. EEUA Handbook No. 33, 1969. Engineering Equipment Users' Association, 20 Grosvenor Place, London SW1. Revised edition, 1971. 45 pp. 15 bibl. refs.

## Sweden

[37] Anvisningar angående skydd mot yrkesfara vid arbete med asbest (Instructions for protection against occupational hazards when using asbestos). Anvisningar No. 52, National Workers' Protection Administration, Stockholm, Sweden. Svensk Reproduktions Aktienbolaget, Stockholm, Sweden, March 1964. 13 pp.